A PRIMER IN BIOLOGICAL DATA

ANALYSIS AND VISUALIZATION USING R

>>>>>>>>>>><<<<<<<<<<

A Primer in Biological Data
Analysis and Visualization Using R

> > > > > > > > > > > > > > > > < < < < < < < < < < < < < < < < < <

Gregg Hartvigsen

COLUMBIA UNIVERSITY PRESS   NEW YORK

Columbia University Press
Publishers Since 1893
New York   Chichester, West Sussex
cup.columbia.edu
Copyright © 2014 Gregg Hartvigsen
All rights reserved

Library of Congress Cataloging-in-Publication Data

Hartvigsen, Gregg.
  A primer in biological data analysis and visualization using R / Gregg Hartvigsen
      p. cm.
  Includes [bibliographical references and index.]
    ISBN 978-0-231-16698-0 (cloth : alk. paper) — ISBN 978-0-231-16699-7 (pbk. :
alk. paper) — ISBN 978-0-231-53704-9 (e-book)

Library of Congress Subject Data and Holding Information can be found on the
Library of Congress Online Catalog.

                                                              2013952140

Cover design: Milenda Nan Ok Lee
Cover image: © Getty Images

References to websites (URLs) were accurate at the time of writing.
Neither the author nor Columbia University Press is responsible for URLs
that may have expired or changed since the manuscript was prepared.

CONTENTS

A PRIMER IN BIOLOGICAL DATA

ANALYSIS AND VISUALIZATION USING R

> > > > > > > > > > < < < < < < < < < <

INTRODUCTION

We face danger whenever information growth outpaces our understanding of how to process it.

(Silver, 2012)

In our effort to understand and predict patterns and processes in biology we usually develop an idea or, more formally, a conceptual model of how our system works. We generally frame our models as testable hypotheses that we challenge with data. As the science of biology has matured our questions of how nature works have gotten more sophisticated and complex. Unfortunately, we are not able to simply look at a table of raw data that we get from an experiment and see an answer to an interesting question with any quantitative level of confidence. Instead, to accomplish this we will learn how to use the R statistical and programming software package to process these data (summarize, analyze, and visualize our results). We also will go a step further and work to understand what these results mean biologically.

Data, graphs, and statistics, oh my! Isn't the interesting stuff in biology really just the cool, living things all around us? It is that stuff but it's *so much more beautiful* when we understand it. Maybe you want to be a vet. Perhaps an early memory for you was loving a little furry thing that purred. However, maybe now you've become a little more concerned about what impact these lovable pets might have on populations of other cute animals that live outside. I recently took a break from writing and looked at an issue of the journal *PLoS ONE* (a well-respected, open-access, online journal). In this journal I saw an article on predation by urban cats in the UK (Thomas et al. (2012)). I "own" three cats and was surprised by the number of prey items that cats brought back to their owners (see Figure 1). It seems that there is a lot of variability in predation rates (the histogram) and that predation rates decrease with increasing urbanization (housing density). Specifically, as seen in the inset graph, the authors state that "There was a significant negative correlation between housing density and annual predation rates on birds ($r = 20.699$, $p = 0.036$)."

When we have questions that we want to answer, such as "what are cats up to when they're outside?," we might read books of fiction, such as the series on Warrior cats (see books by Erin Hunter, which is actually a pseudonym!). In biology, however, we seek to understand things like cats by collecting, interpreting, analyzing, and visualizing data. This book is designed to help you to be able to do this. If you're interested in other disciplines I hope the examples in this book help you, too! I also hope that as you use this book you lose any fear you might have of data and instead seek out and work with data and understand what they tell you about the things that got you interested in biology in the first place, like cats (or, more likely, dogs).

WHAT THIS BOOK IS (AND ISN'T)

This book is designed to help you collect, organize, analyze, and visualize data. I assume you have not heard of the free, open-source program R and I will, therefore, introduce you to how to use it to accomplish these goals. Although I imagine you have had some experience making graphs and calculating a few descriptive statistics (e.g., mean and standard deviation in Excel) I assume you haven't done this. If you don't know Excel, or don't have access to it, you will be able to do all the heavy lifting in this book. I assume you have not taken a course in statistics.

This book, therefore, aims to give you a foundation upon which to become a better student of science and a better consumer of scientific information. More specifically you will learn how to

- formulate hypotheses,
- design better experiments,
- do many standard statistical procedures,
- interpret your results,
- create publication-quality visualizations of your results,
- find help so you can solve your own problems, and
- write a simple computer program.

You shouldn't expect to read this book and become a quantitative guru. Instead, you should hope to become competent at finding answers to some

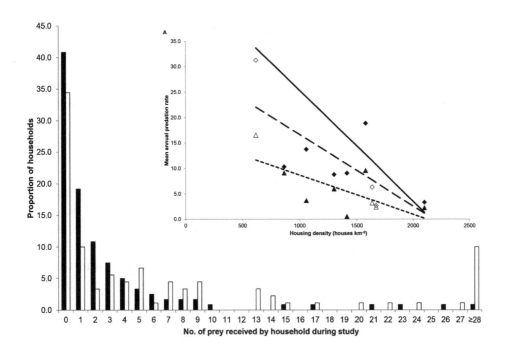

Figure 1: Two figures from a recent paper on urban cat predation rates (Thomas et al. [2012]). The larger graph is a histogram showing percentages (instead of the usual frequencies, or counts) for the number of prey returned to households. Black and white bars are for households with a single-cat versus multiple-cats, respectively. The insert is a scatterplot with best-fit straight lines added for birds, mammals, and for both animal groups combined. The combined data points have been omitted! The relationships are analyzed and discussed in the paper as "correlations" and, therefore, adding lines is inappropriate (see the box on page 138). The graphs and resulting analyses were likely done using R, but that doesn't mean they are correct! After you work through this introduction you should be able to comfortably assess these data, correctly perform the analyses and create more appropriate visualizations.

of your questions, such as "are these two samples different?" and "is there a significant linear relationship between my variables?" You will become a resource to the people around you. And if you put in some time playing with R you will be the go-to person for data.

I have written this book primarily with the hope that you'll feel more comfortable with complex biological problems. It has grown out of what I have seen challenge my own undergraduate students. But it also covers some topics that I think are fun and valuable to know how to do (e.g., programming). The chapters end with problem sets for you to challenge yourself to use what you have learned. Some of the data are real while some are merely *realistic*. I also have included solutions to the odd-numbered problems at the end of the book. Finally, the book is filled with R code. You should type this is in yourself because this helps with the learning process. You can, however, go to `https://github.com/GreggHartvigsen/PrimerBiostats` and download all the code from this book.

This book is neither a formal introduction to R nor a statistics textbook. Instead, this book helps you to you solve problems you're likely to encounter in your undergraduate program in biology. I work to explain what statistics are and how to share and interpret scientific results. After working through this book you should be able to solve a variety of problems with the most widely used statistical and programming environment. I hope you will no longer be afraid of data and will be more able to enter data into the computer, test hypotheses, and present your findings.

So, this book should help you make more appropriate and professional, scientific visualizations and discover findings that might have otherwise been missed. You will no longer be satisfied with hearing from anyone things like "Well, it looks significant" or "there seems to be a trend in the data." So, for the rest of your career, I hope you become the person who says "We can test that! Let me get my laptop."

WHO REALLY NEEDS THIS?

In this book I work not only to present visualization and analytical techniques but to explain why we do all this. There's an unfortunate misconception that we don't really need all this quantitative stuff in biology. I have heard several times the following line of thinking:

> Why do we need to use statistics in biology? If the hypothesis is clear, the experiment is designed correctly, and the data are carefully collected, anyone should be able to just look at the data and clearly see whether or not the hypothesis is supported. Statistical procedures are simply safety nets for sloppy science.

As you work your way through this book you'll see why the above thinking limits scientific exploration, understanding, and the ability to make predictions about natural phenomena. Here is a brief list of reasons why statistics, mathematics, and appropriate visualizations are critical for understanding biological systems:

1. Statistical procedures help us determine whether data are consistent with hypotheses. Data from modern biological experiments are unable to "speak for themselves." Data, instead, require rigorous evaluation, which is appropriate because they are often hard to collect. Statements based on opinion, such as "I don't believe global warming is happening" or "I believe this drug will cure cancer," fall outside the realm of science.

2. Based on our results from data analyses we often develop formal mathematical models that help us to understand and explain how systems work. We do this by developing quantitative predictions that we assess with data.

3. Biologists often work to understand how multiple factors work together, often in complex, non-linear ways, to affect biological systems. To determine the individual effects and the combined interactive effects we need to develop and conduct complex experiments to illuminate biological patterns and mechanisms that cause these patterns. We then use sophisticated data analysis procedures and visualization techniques to answer today's challenging questions.

Biology is one of the more complex sciences. I will admit that, at times, some questions can be pretty simple. Imagine, for instance, that we have 100 randomly selected pea pods and expect a 3:1 phenotypic ratio of yellow to green peas. We should expect to see a ratio of 75 to 25 yellow to green peas. We, however, are unlikely to see exactly this ratio. If, instead, we find a ratio of 78:22 we can see immediately (without statistics!) that this is not a 3:1 ratio. Are you prepared, based on this finding, to conclude that this system does not follow the well established rules of segregation? Scientists are predisposed by their profession to be skeptical and, therefore, will not accept a statement like "Trust me that our finding of a 78:22 ratio demonstrates that Mendel was wrong!"

Our goal is to understand biological systems. Unfortunately, anything interesting nowadays is complex (even determining if our data adhere to a simple 3:1 ratio!). With quantitative tools we can better understand how natural systems work. Only then might we be able to make accurate and useful predictions. Science relies on a strong foundation of statistics, mathematics, and the visualization of results, all of which are available to you through the R statistical and programming environment.

ADDITIONAL RESOURCES

There are far too many great sources of information on data analysis, statistics, visualizing information, and programming to list them all here. This book is a very basic introduction to all of these topics. I hope you seek more information in all of these areas. If you do, here are a few recommendations that go more deeply into different subsets of the topics covered in this book:

General introductions to R

1. An introduction to R. Venables and Smith (2009)
2. A beginner's guide to R. Zuur et al. (2009)
3. R for dummies. Meys and de Vries (2012)
4. The R book. Crawley (2012)
5. R in a nutshell: A desktop quick reference. Adler (2012)

Statistics books

1. A primer of ecological statistics. Gotelli and Ellison (2012)
2. Statistical methods. Snedecor and Cochran (1989)
3. Biostatistical analysis. Zar (2009)

Statistics books specifically using R

1. Introductory statistics: a conceptual approach using R.   Ware et al. (2012)
2. Foundations and applications of statistics: an introduction using R. Pruim (2011)
3. Probability and statistics with R. Ugarte et al. (2008)

Visualization using R

1. ggplot2: elegant graphics for data analysis. Wickham (2009)
2. R graphics cookbook. Chang (2013)

Programming using R

1. The art of R programming. Matloff (2011)
2. http://manuals.bioinformatics.ucr.edu/home/programming-in-r

CHAPTER 1

INTRODUCING OUR SOFTWARE TEAM

In science we are interested in understanding systems that are complicated. Our use of quantitative approaches gives us the ability to not only understand these systems but also to predict how a system might behave in the future (or maybe even how it behaved in the past). As we work to understand and predict complex biological systems we need computational help. You probably have written lab reports using only a calculator. This should be avoided for a variety of important reasons:

1. Difficulty in verifying that you entered the data correctly. (I *think* the numbers are right.)
2. Difficulty in repeating the analysis. (I'm not doing it again because I might get a different answer!)
3. Inability to share your analytical approaches and results. (Sorry, I hit the all-clear button! You have to trust me.)
4. Inflexibility in how the data are analyzed. (You wanted me to do what?).
5. Inability to make and share appropriate graphs. (Can I take a picture of the graph on my calculator with my phone and incorporate that in my lab report?)

To solve these shortcomings we will use Excel and R.

You may be somewhat familiar with Excel but probably have little or no experience with R. Therefore, I welcome you to the world of R! I know this might be a scary place for you at first. I bet R is really different from all the programs you've used. Fortunately, this introduction is intended for newcomers. But as you proceed you will learn how to do some really amazing things with R. You'll gain independence with practice. R is like playing an instrument, a sport, or learning a foreign language—they all require practice. I have confidence that you are capable of using R to solve interesting problems. And the more time you spend at it the better you will get.

1.1 Solving problems with Excel and R

For many analytical problems we will be able to use just R. However, in biology, we often test our ideas, or hypotheses, with large amounts of data. We, therefore, will try to use Excel for what it does well (allows us to enter and organize our data). But we will not use Excel to do what it doesn't do well (statistical analyses, modeling, and visualizing data). Instead, these core scientific skills are best done with R. If you love Excel then you'll be happy to know we're not abandoning it—Excel has its place.

It is important to recognize that doing things well is rarely easy. Writing a good poem, playing tennis well, or doing ballet well are all hard. And conducting hypothesis tests correctly and making professional-quality graphs are not simple, one-click operations.

At first you will likely think that making graphs and performing statistical tests in R are absolute nightmares. (And when you become a skilled R programmer you'll still be challenged at times!) But the days of skipping an analysis or accepting a ugly or incorrect graph because "that's the best I can do with Excel" are over. You can do it in R! Therefore, in this introduction we will discuss Excel but focus mainly on R. It is the combination of using Excel to organize our data and R for analyses and visualizations that will allow you to ask and answer questions in biology.

You still may be wondering why you can't just do this all in Excel. Here is a sampling of reasons why R is clearly better than Excel for problem solving in biology. With R you can:

1. create professional, publication-quality visualizations;
2. conduct quantitative analyses, both analytical and statistical (e.g., do a t-test, solve systems of differential equations, conduct non-linear regression, use matrix algebra, conduct signal processing, perform wavelet analysis, analyze fMRI data, do genome analyses, and create phylogenetic reconstructions, to name a few);
3. build statistical tests that can be repeated easily and shared with anyone. These tests might rely on their own data, data read from a file, or data acquired directly from a website;

4. do the same thing and work the same way on computers running Mac, Windows, and Linux;

5. write computer programs, such as modeling a population growing over time, using an object-oriented language;

6. access modern analytical tools for biologists that are being developed right now, right here, and no where else;

7. use and receive widely available help from the R open-source community;

8. use open-source software that provides solutions that are "auditable," meaning you can understand and explain to others how you got your results (there are no black boxes - it's open software!);

9. write a document like this. This environment allows one to compile together in one document words, mathematical equations, computer code, statistical tests and output, and professional-quality graphs, all within the free, open-source LaTeX typesetting environment;

10. carry a research project, paper, all the data, AND carry the entire software package for doing the analysis on a low-capacity flash drive;

11. rest assured that your investment in skill building will pay off well into the future. You don't have to hope you'll have access to the program when you move on to your next stage of life (which could be in a hospital in Ghana!);

12. enjoy these benefits because open-source means R is free!

Your ability to use R to make informed, evidence-based conclusions likely will provide you the most valuable set of skills you'll learn as an undergraduate science major. If you keep this skill set you will be highly marketable. R helps you speak the language of science, which is written in mathematics, statistics, and data evaluation and visualization. This ability to answer scientific questions and present your results professionally is finally in your hands.

Your ability to use R helps fulfill an important goal that was synthesized in the report *Scientific Foundations for Future Physicians* produced by the American Association of American Medical Colleges and the Howard Hughes Medical Institute, 2009. The authors of this report downplay the importance of memorizing facts and, instead, encourage students to learn to

> apply quantitative reasoning and appropriate mathematics to describe or explain phenomena in the natural world.

Additionally, in the report *Vision and Change in Undergraduate Biology: A Call to Action,* produced jointly by the American Association for the Advancement of Science and the National Science Foundation (2009), six "core competencies" are advocated for undergraduates in biology. Below are four of the six competencies that are directly addressed in this book:

- Competency #1: ability to apply the process of science (understand how to formulate and test hypotheses);
- Competency #2: ability to use quantitative reasoning (e.g., use statistics and quantitative modeling approaches);
- Competency #3: ability to use modeling and simulation;
- Competency #5: ability to communicate and collaborate across disciplines.

The reason you have this book is to achieve these goals. So, it's time for us to get going.

1.2 INSTALL R AND RSTUDIO

We're going to make the installation of your R environment a two-phase process. First we will install R, which is a basic program with a simple interface. You can do everything discussed in this book in this environment. Consider this the engine, frame, wheels, and steering wheel for a car. It'll get you to wherever you want to go. The second step is to install RStudio, which makes it a much more comfortable ride. For both of these you can simply accept the defaults offered by the programs during installation.

1. **Install R**. In a web browser, search simply for the letter "r," or go to http://cran.r-project.org/. Follow the instructions to install the correct version of R on your computer. Note that if you borrow a computer but you don't have the proper administrative rights you usually can install R on the computer's desktop. If you have a Mac running the OS prior to version 10.6 then the latest version of R may not run. Check out the information at http://cran.r-project.org/bin/macosx/ if your installation doesn't work.

2. **Install RStudio**. Back in your web browser, search for "RStudio" or go to http://rstudio.org/. Follow the instructions to install the correct version of RStudio on your computer. Again, the most recent version of RStudio does not seem to work on Mac OS prior to 10.6. For Mac OS 10.5 you can download RStudio at https://s3.amazonaws.com/rstudio-dailybuilds/RStudio-0.95.265.dmg.

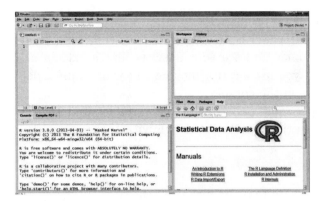

Figure 1.1: Screen shot of RStudio running on the Windows operating system.

Once you have RStudio running, it should look much like the screenshot from a Windows-based computer (Figure 1.1) or a Mac-based computer (Figure 1.2. You may see only one large window on the left side. If you do you might click on `File -> New` and open a new script file or document.

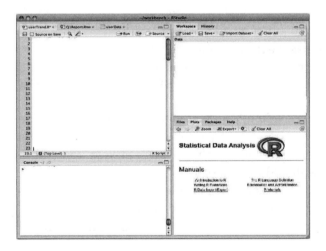

Figure 1.2: Screen shot of RStudio running on the Mac operating system.

RStudio should show four panels. The upper-left panel is where you can enter lines of code into a "script" file. RStudio allows many script files to be open and uses tabs to help you keep track of them. You'll find out more about scripts in section 1.5 on page 19.

The lower-left panel is the "console" (or "command prompt") where you can type commands and see your answers, like a calculator (see section 1.4 below). Also, if you use a script file and run your commands, the text output will appear here in the console.

In the upper-right panel you'll see variables and their values after you declare them. In the console, set a = 5 with this command:

```
> a = 5
```

[Note that you should not type the ">" character.] The a and 5 should show up in the upper-right panel. Finally, the lower-right panel displays help information and plots you create.

1.3 GETTING HELP WITH R

If you like to get help when using a program you're in luck. There are many ways to get help with R and RStudio. If you don't know how to do a t-test or an ANOVA in R then you can search within RStudio or you might just search on the web for help.

You can get help in the console like this:

```
> ?mean # Gives help on function mean()
> ??mean # Finds all occurrences of "mean" in the help system
> ?"if" # Get help on the control keyword "if" (same for "for")
```

Do not type the leading ">" character—R provides that for you by default in the console. Also, text that follows the # sign is ignored by R. These "comments" can help you, or other readers of your code, to understand what the commands should accomplish.

You also can get help from inside your script files (upper-left panel) by placing the cursor on a keyword and hitting the "tab" key. A helpful pop-up window should appear with help on your function or similarly spelled functions.

Another important and rich source of help is available online. Feel free to explore this by simply performing a search in a web browser, such as "r" and "mean". For more help you can go to one of these sites:

- http://cran.r-project.org/doc/manuals/R-intro.html
- http://www.statmethods.net/

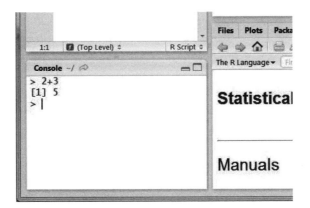

Figure 1.3: Adding two numbers in the console of RStudio.

- http://rseek.org/

or check out the section on where to go from here in section 13.1 on page 206.

1.4 R AS A GRAPHING CALCULATOR

Assuming you have installed R and RStudio on your computer, start RStudio. Note that from here on when I refer to R I will assume you'll be using the RStudio interface.

Let's begin by running a few commands in the console. Click inside the console (lower-left panel) to activate it and try adding 2 and 3 (see Figure 1.3). Finish by hitting the <enter> key:

```
> 2 + 3
```

```
[1] 5
```

Did you get 5 (see Figure 1.3)? You have just run a command at the command prompt. You have used your computer to compute! So far I hope this has not been painful!

The [1] before the answer seems a bit strange. R is actually reporting that 5 is the first number in a possible array, or vector, of numbers. Sometimes your answer will have lots of values and R will provide you with this counter to help you find values, but more on that later.

Below is a variety of calculations for you to try. The more you play the easier this will be. You should check that you get the same answers as I did.

```
> 5-1 # Subtraction. Did you get 4?

[1] 4

> 2*3 # Multiplication

[1] 6

> 7/3 # Division

[1] 2.333333

> sqrt(9) # Use the sqrt() function to get the square root of 9

[1] 3

> 9^2 # 9 squared

[1] 81

> log(3) # Natural logarithm of 3

[1] 1.098612

> log(3,10) # Log of 3 (base 10), or use log10(3)

[1] 0.4771213

> exp(3) # This is e^3

[1] 20.08554
```

What if a command doesn't work? R is really picky in how you enter commands. There's a little wiggle room with spaces, for instance (5+3 and 5 + 3 both work), but R is very finicky about non-space characters (e.g., sqrt{9} doesn't work). R will give you an error message if something's not quite right:

```
> sqrt{9} # WRONG! Need parentheses (), not curly braces {}
Error: unexpected '{' in "sqrt{"
```

Computers, and programs like R, generally do exactly what you tell them to do, which might not be what you intended them to do! If something goes wrong R will return an "error message" that should be somewhat helpful, as you saw above, but it's never very friendly about it. It's important not to take this personally—for R it's all business.

Let's try some more complicated calculations. The follow lines of code rely on you providing some data in an "array," which is a single variable with a bunch of similar objects, like numbers or words, packaged together. Once the data are in an array then far more interesting things can happen. Be sure to try these because we'll see arrays throughout the rest of the book.

```
> 1:5 # create an array of integers from 1 to 5

[1] 1 2 3 4 5
```

We can use the combine function (c()) to group any set of numbers together into an array.

```
> c(1, 2.5, 3, 4, 3.5) # combining five numbers into an array

[1] 1.0 2.5 3.0 4.0 3.5
```

We can even store those numbers in a "variable" so that we can use them again. Here's how to store the numbers into the variable a:

```
> a = c(1, 2.5, 3, 4, 3.5) # store numbers the variable "a"
```

Storing numbers in an array is very common in statistics. We'll learn more about variables when we discuss data in chapter 2. Now that the data are in an array we can perform a variety of operations on them.

```
> sum(a) # sum up all values in array "a"

[1] 14

> length(a) # tells you how many numbers are in "a"

[1] 5

> sum(a)/length(a) # this calculates the mean
```

```
[1] 2.8

> summary(a) # more descriptive stats for "a"

   Min. 1st Qu.  Median   Mean 3rd Qu.   Max.
    1.0     2.5     3.0    2.8     3.5    4.0

> a/5 # divide each value in "a" by 5

[1] 0.2 0.5 0.6 0.8 0.7

> a[5] # returns the fifth element in array "a"

[1] 3.5

> a[1:3] # returns the first three elements of array "a"

[1] 1.0 2.5 3.0
```

Sometimes you will need to make a sequence of numbers. We can do that using the **seq()** function, instead of typing them all into the computer. The first line prints the sequence to the screen. The second line stores the sequence in a variable called **my.seq**.

```
> seq(1,10, by = 0.5) # sequence from 1 to 10 by 0.5

 [1]  1.0  1.5  2.0  2.5  3.0  3.5  4.0  4.5  5.0  5.5  6.0
[12]  6.5  7.0  7.5  8.0  8.5  9.0  9.5 10.0

> my.seq = seq(1,10, by = 0.5) # store result in variable "my.seq"
```

We can now see that the numbers in square brackets represent the index number in arrays. Above, the 12th number is 6.5. The 13th number is 7.0. Notice, too, that indexing in R begins with [1], which differs from some other programming languages, such as C, which starts with zero.

Sometimes, instead of a sequence of numbers, we need to repeat numbers, or even letters, for different experimental designs. Let's try this using the repeat function (**rep()**):

```
> rep(c("A","B","C"), times = 2) # entire array twice

[1] "A" "B" "C" "A" "B" "C"
```

```
> rep(c("A","B","C"), each = 2) # each element twice
```

```
[1] "A" "A" "B" "B" "C" "C"
```

You also can combine calculations within the declaration of an array.

```
> p = c(1/2,1/4,1/4) # three proportions saved in an array
> sum(p) # p should add to 1.0
```

```
[1] 1
```

R also is good at making graphs. Sometimes you need to see what a sine curve looks like or, perhaps, a simple polynomial. Suppose you are asked (or simply want) to view a function such as this:

$$y = 2x^2 + 4x - 7 \tag{1.1}$$

over the range $-10 \le x \le 10$. In R it's really easy! You can use the curve() function (see Figure 1.4):

```
> curve(2*x^2 + 4*x - 7,-10,10)
```

The curve() function requires us to provide at least three "arguments," each separated by a comma (we can provide more if we want to enhance the graph; see Box 5.1 on page 68). The first thing required is the equation we want R to graph. There must be an "x" in your equation and no other unknown variables or R will throw an error message to the console. Additionally, the function needs a starting value for x and an ending value for x. If entered correctly R will create a smooth curve over this range (Figure 1.4). We will use the curve() function to help us solve some tough problems later in the book (e.g., section 11.5 on page 171).

Here's one more thing to try. Histograms are great graphs, as we'll see, to visualize the distribution of data. The elusive "bell-shaped" curve of the "normal distribution" can be made using data drawn from the standard normal distribution ($\bar{x} = 0, s = 1$). You just need to send them to the hist() function to get a nice graph of them.

```
> hist(rnorm(10000))
```

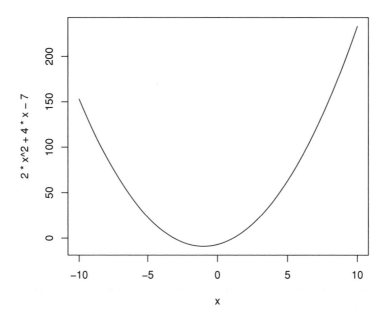

Figure 1.4: The graph of the function $y = 2x^2 + 4x - 7$ over the range $-10 \leq x \leq 10$, made using the `curve()` function.

I didn't reproduce the graph because we'll work on this later (see chapter 5). But if you typed that correctly at the command prompt and hit `<enter>` I hope you saw a cool "histogram." What's happening under the hood is the following. First, `rnorm()` is a "function." Functions are routines that do a bunch of stuff behind the scenes (see Box 1.1). They "take" arguments (the stuff you include in the parentheses) and then "return" stuff. In the `rnorm(10000)` call above we send the number 10,000 (without the comma) to the `rnorm()` function as a single "argument." The function `rnorm()` then "returns" an array (a "variable" with a single row or column of numbers) of 10,000 random numbers, drawn from a standard normal distribution. The call above then uses those 10,000 values as the "argument" to the `hist()` function. The `hist()` function then divides the numbers into bins and, behind the scenes, calls the `plot()` function that makes the histogram, displayed in the graphics window. There's a lot going on with only that one, simple line of code!

> **Box 1.1. Functions.** In programming languages, like R, a *function* is a previously defined set of instructions that does something. A function may or may not require "arguments" as input. Multiple arguments need to be separated by commas. Functions may return a variable and/or perform an operation. The `mean()` function, for instance, takes an array of numbers and returns the arithmetic mean. You can choose to send the `quit()` function no argument and it returns nothing (but RStudio will be closed!). R has many functions to simplify your work and you can even write your own functions, too (see section 11.1 on page 161).

1.5 Using script files

You can do everything you want at the command prompt. But for most jobs it is best to create a "script" file. This is a plain text file, created from within RStudio, that contains a series of commands that can be run either one at a time or all together as a "batch" job. You should save script files in a folder for your current project. In that folder you might have associated data files (e.g., Excel files), a presentation file, and maybe a lab report or paper. When you return to a project you open the script file, like any document from other programs, and get to work. Be sure to give the files and folders meaningful names and save your script file frequently. You can see what a typical project, such as a laboratory exercise, looks like in R in Box 1.2.

To run a line of code in a script file you should have it open in RStudio. Place the cursor on the line you want to run and hit `<ctrl><enter>` (Windows PC) or `<cmd><enter>` (Mac). You also can highlight any chunk of code, like you might do in other programs, and run only that which you have highlighted. I will often want to run the entire file and so hit `<ctrl><a>`, which highlights all the text in the current document, then `<ctrl><enter>` (replace `<ctrl>` with `<cmd>` on a Mac).

It is time to create a new script file and save it with a meaningful name in a practice project folder. Type the following line of code into it, save the file, and run the line. Note that the structure and R code have to be correct for it to work. If you forget a comma or spell the `sqrt()` function `SQRT()`, R will provide you a mildly helpful error message. You also should include comments to yourself (after the `#`). These "comments" are ignored by R.

```
cat("The square root of 2 is", sqrt(2),"\n") # square root of 2
```

Box 1.2. What does a typical R project look like? Let's imagine that you're in a laboratory class and several groups collected and pooled their data. The data have been made available in an Excel spreadsheet. Here are the steps you might take to complete the lab:

1. Download the data file and store it in a folder for this class/lab/project.
2. Open RStudio and create a new script file, give it a meaningful name, and save it in the project folder with the data.
3. Set the working directory in RStudio to the location of your script file (Session -> Set Working Directory -> To Source File Location).
4. Write yourself some commented text (lines that start with the # symbol) about the project so when you return you'll know what you were trying to do.
5. Write lines of code that do the following:
   (a) Read in the data from the spreadsheet file (see section 2.2 on page 27).
   (b) Explore your data. For example, you will likely gather summary statistics (see chapter 4 on page 47), test whether your data are normally distributed (see section 4.4 on page 56), and/or create visualizations (see chapter 5 on page 67).
   (c) Conduct the appropriate statistical test(s) (see chapters 7–11).
6. Once you've completed the tasks above you'll need to copy your graphs and statistical output from RStudio to your laboratory report.

If you run this line then you should see pretty much what I see (Figure 1.5). Organizing your code into script files and placing those into well-named project folders will *greatly* simplify your life. Be organized from the start—it will save you time. One last beauty of RStudio is that when you close the program, and then later open it up, it will reopen the files that you left open in RStudio when you last closed it.

1.6 EXTENSIBILITY

When you install R and RStudio you will have the basic version of the software. As you progress in your use of R you'll likely need to add more tools (or get the latest version of your existing tools). To get those tools you can install

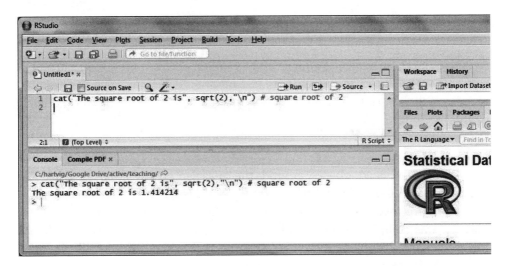

Figure 1.5: RStudio after running the script file that is in the upper-left panel. The output is in the console (lower-left panel).

packages from Internet servers located around the world. Once these packages are installed you have to "load" them with the `library()` command at the beginning of your session.

At first this probably makes R seem like an incomplete software package. It's actually a very efficient use of your computer and its resources. You only install what you need. The major computer packages with which you are familiar require a lot of time to install and consume a lot of hard drive space. R is much smaller, faster, and customized by you for what you do.

The following four lines will install the four packages that this introduction assumes you have. When you try to install a package for the first time during an RStudio session, you will be asked to choose a "mirror" website, so you need to be connected to the Internet. You should choose one that is geographically nearby (although, at times, I'll go international for fun, e.g., using a mirror in Iran). Once these packages are installed you can use them without being connected to the Internet. You also can save the packages to a flash drive if you're at an Internet cafe in an exotic study abroad location and install them later. Here are the packages I will assume you have installed:

```
> install.packages("e1071")
> install.packages("deSolve")
> install.packages("plotrix")
```

```
> install.packages("UsingR")
```

To make these packages that you've installed active during your current R session, you need to use the `library()` function. If, for instance, you wanted to solve a differential equation then you need to include this in your code:

```
> library(deSolve)
```

[Note that you need quotes around the package name when you use the `install.packages()` function but not in the `library()` function.]

If you are using an older version of R then you will be warned that the package was compiled using a more recent version of R. This warning is not likely to cause a problem. If you get this message you should install the latest version of R, at your convenience. Note that "warnings" are just warnings. You still might be fine. An "error message," however, means something is wrong and you'll have to fix it.

Now that you've installed R and RStudio, used it as a graphing calculator, and installed a few packages, it's time to get to work. Below are a few problems for you to try.

1.7 PROBLEMS

1. Find the solutions to the following problems:
   - (a) $\sqrt{17}$
   - (b) $\log_8(10)$ # read the help on the `log()` function (> `?log`)
   - (c) $17 + (5x + 7)/2$, if $x = 3$

2. What's wrong with the following statement?

   ```
   > e^10
   ```

   Enter it into R to see the error message returned. Fix the problem and determine the value of e^{10}.

3. Find the "basal area" of a tree (cross-sectional area of a tree trunk) if the diameter is 13.5 cm. Hint: area $= \pi \, r^2$.

4. What is the volume of the Earth in km^3 if the radius is 6367 km and we assume it is a perfect sphere? Check your answer by searching online.

5. The following are masses for chickadee chicks from several nests, measured in grams:

3.2, 6.7, 5.5, 3.1, 4.2, 7.3, 6.0, 8.8, 5.8, 4.6.

(a) Combine the data into a single variable array called `my.dat`.
(b) What is the arithmetic mean mass of the chicks?
(c) What is the standard deviation?
(d) What is the variance?

6. Create an array of numbers from 0 to 5, by steps of 0.1 using the function `seq()`. Save the result in a variable called `S`.

7. You are riding in your friend's monster truck and you've stopped yet again at the gas station to fill up. You wonder what the gas mileage is for this vehicle. Your friend tells you it went 230 miles on the 28 gallons of gas, the amount just put in the tank.

(a) What's the fuel efficiency for this vehicle in miles per gallon (mpg)?
(b) If the tank holds 32 gallons how far can this vehicle be driven?

8. A package with a lot of data is called "MASS." Install this package, load it, and look at the dataset called "Animals." Remember that variable names are case sensitive. Just looking at these data, what do you think the relationship is and should be?

9. The Michaelis-Menton curve is used in enzyme kinetics. It can be represented by this equation:

$$v = \frac{V_{max} \cdot [S]}{K_m + [S]}$$

where v is the velocity of the reaction, V_{max} is the maximum velocity, $[S]$ is the concentration of the substrate, and K_m is the Michaelis-Menton constant. Create a curve (using the `curve()` function) of this relationship for concentrations of $[S]$ from 0–5, assuming that $V_{max} = 0.9$ and $K_m = 0.12$.

10. Install, if you haven't, and load the `UsingR` package. Issue the following command and the console:

```
> data(package = "UsingR")
```

This lists all the datasets available in this package. The last dataset contains catch data for yellowfin tuna. You can graph the data with the simple command:

```
> plot(yellowfin)
```

What is the average number of yellowfin tuna caught in the last five years of the data (1999–2003)? Use R to count the number of years of data and then use the method of extracting specific data points from an array (remember that you can do something like my.dat[3-5] to get the 3$^{\rm rd}$ to 5$^{\rm th}$ data points in an array called my.dat.

∧
3:5

CHAPTER 2

GETTING DATA INTO R

There are many ways to get data into R and which technique you use depends on a variety of factors. All data are eventually stored in "variables" in R. A variable is a named object in R that is used to store and reference information (see Box 2.1). The method we use to do this usually depends on the amount of data we have. The three basic approaches are as follows:

1. fewer than about 25 values: create a variable and assign data to that variable using the `c()` function;
2. more than about 25 values: enter data into Excel, save the file in a `.csv` format, and then read the data file into R (`read.csv()`); or
3. read data in from a website (`read.table()`).

We'll discuss each of these methods in more detail in the coming sections. Note that getting data into R accurately is a critical step in understanding what the data tell us. Extreme care needs to be taken in this step or all our efforts in doing the work could be lost. It's always a good idea to have someone help you with double-checking that the data have been entered into the computer correctly.

2.1 USING c() FOR SMALL DATASETS

The best way to get small amounts of data into R is to use the combine function (`c()`) in a script file. This function, as it sounds, groups together data into a single "array." Open a new script file and create your variables and assign the data. For example, if we imagine we have heights for each of four men and women then we can do the following:

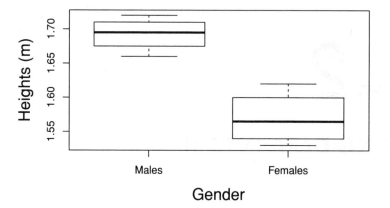

Figure 2.1: The heights of males and females, presented using the `boxplot()` function.

```
> males = c(1.72, 1.69, 1.70, 1.66) # heights in meters
> females = c(1.55, 1.62, 1.58, 1.53)
```

After running these lines our data are available using these two variables by name. We can do things like get the mean or median of the heights:

```
> mean(males) # calculates the arithmetic mean

[1] 1.6925

> median(females) # calculates the median for continuous data

[1] 1.565
```

or plot the data by sending the arrays to the `boxplot()` function (see Figure 2.1).

```
> boxplot(males,females, names = c("Males","Females"),
+          xlab = "Gender",ylab = "Heights (m)", cex.lab = 1.5)
```

We will discuss visualizations in detail in chapter 5.

2.2 READING DATA FROM AN EXCEL SPREADSHEET

If you have data with more than about 25 values you should probably enter these first into a spreadsheet and then read that file into R. Programs like Excel are great at helping you organize your data and allowing you to easily double-check that you entered the values correctly.

Box 2.1. What is a "variable"? Variables are words (also called "objects" in R) that store information. We usually use them to store a number or a group of numbers for later use:

```
> my.height = 68 # height in inches
> my.height

[1] 68
```

or they might store characters (e.g., "**Gettysburg**" could store the Gettysburg Address):

```
> Gettysburg = "Four score and seven years ago...."
> Gettysburg

[1] "Four score and seven years ago...."
```

Naming conventions differ but you should consider creating names that are as short and descriptive as possible (but no shorter!). This becomes particularly important if you need or want to share your code with others. It's also important if you're working on a project and need to return to the project in a week or more (sometimes we forget!). Good names and commented code can help us save time. If, for example, you have the masses for five dogs you might use multiple words, separated by periods, for example:

```
> dog.mass = c(4.3, 6.7, 5.2, 3.7, 4.4) # mass of our 5 dogs
> dog.mass # view and double-check the data

[1] 4.3 6.7 5.2 3.7 4.4
```

You also may find data from sources such as the Centers for Disease Control (CDC) that allow you to download files to your computer in a text format; usually a "comma-separated variable" file ending with .csv. R can read native Excel spreadsheets (e.g., in .xlsx format) but it is best if you save the data in the comma-separated variable format (.csv). The .csv format is highly portable and likely will be used for decades. Finally, within R you can combine datasets and create new spreadsheets that you can save to disk and share with colleagues.

When you create a spreadsheet of data follow these rules:

1. Use descriptive, one-word (or at least no spaces), unique column headings in row 1. These should start with letters but may contain numbers (see Figure 2.2). Note that these will be the names of your variables.

2. Begin the data on row 2. Do not include comments or empty cells in the middle of your data, if at all possible. Variables should start with letters. Variable names can have numbers (e.g., "trmt5"). Numbers (your data) should not have characters. As with other statistics programs if you have a column with thousands of numbers and a single word the entire column will be considered text (you won't be able to graph the numbers or do the usual statistics on them)

3. Save the file as a text file in .csv format. Note that Excel formulas will be replaced by the values. If you don't want to lose those then save the file as an .xlsx (native) Excel file and then save a copy of it in the .csv format. The .csv format will save only the first sheet of a spreadsheet file. Avoid using multiple sheets within a single file (otherwise, you can use the xlsx package to read in different sheets from an Excel file). You can use text formats other than .csv (e.g., tab-delimited files) but you will need to use the read.table() function and specify the character used to separate your data values (e.g., sep = "\t" for tabbed data).

4. In RStudio change the active directory to where you saved the file (Session -> Set Working Directory -> Choose Directory).

5. After choosing the directory you can check that your file is in the current directory. In the console type:

```
> dir()
```

If you see your data file you're in business. If not, set the working directory to where the file is or move your file to where you want it. You can, instead, click on the "Files" tab in the lower-right panel of RStudio.

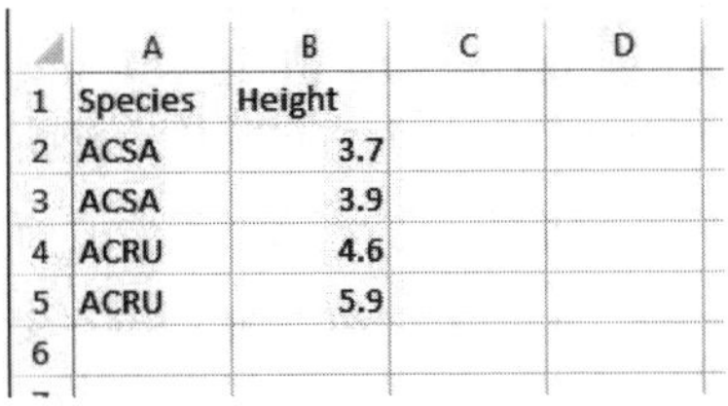

	A	B	C	D
1	Species	Height		
2	ACSA	3.7		
3	ACSA	3.9		
4	ACRU	4.6		
5	ACRU	5.9		
6				

Figure 2.2: A sample spreadsheet with data.

This allows you to look for files through the file structure and click on files (e.g., a `.csv` file) and view it in RStudio.

6. Read in the file, storing it in a "dataframe." Below I read in a file (e.g., "filename.csv") and store it in the dataframe `my.data`:

```
> my.data = read.csv("filename.csv")
```

The filename must be in quotes. The variable `my.data` can be any legal variable name (names can't start with numbers and shouldn't have special characters).

7. Check that the file was read in correctly and that the column headers (variables) are contained within your file:

```
> names(my.data) # should return the first row names
```

You should see the variable names which came from the first row of your data file. These may be different than you intended because R changes them a bit to be legal (e.g., spaces are converted into periods). For an overview of this important process see Box 2.2).

Once you are done with a variable you can remove it from the current environment with the `rm()` function. This can be a good idea if you are going to move on to another project without closing down RStudio. To remove the variable `x` you simply type the following `> rm(x)`. You can remove all variables by clicking on the "Clear All" button in the upper right "Workspace" panel of RStudio or by typing in the console `> rm(list=ls())`.

Box 2.2. Reading data into R If you need to read in data that are stored in a file here's what needs to happen:

1. The data file needs to be saved somewhere on your computer, preferably as a `.csv` text file. It's an optional file type in Excel, found if you choose "Save As...' in Excel.
2. Open RStudio and tell it the location of the file (set working directory).
3. Read the data in and store them in a variable, e.g.:

```
> my.data = read.csv("file.csv")
```

4. Your variable (`my.data`) is a dataframe that holds all the data from the first sheet of the `.csv` file. You access the data in R using that variable name. The dataframe name alone points to all of the data. You can access individual columns of data using the $ sign convention (e.g., `my.data$height`). Here's what this might look like:

```
> my.data = read.csv("plant data.csv")
> my.data$heights # print heights to screen
> mean(my.data$heights)
```

We discuss dataframes more in chapter 3 on page 36.

If you need to change a value in your data you should change it in the original `.csv` file. You can change it in RStudio, but if you start over then you'll read in the original file with the wrong data. You should detach the data from within RStudio (e.g., `> detach(my.data)`), make your changes in Excel, save it, then read it back into RStudio. If your data file is called `my.data` then you can edit these data by clicking on the variable name in the upper-right panel in RStudio and change individual values. These changes are *not* saved permanently in the data file unless you choose to save the dataframe to disk with the `write.csv()` function (e.g., `>write.csv(my.data,"newfilename.csv")`).

2.3 READING DATA FROM A WEBSITE

We often find that we need data that are posted on a website. Reading the data directly from websites is appropriate, especially when the data are continuously being updated. We can, for example, get the number of sunspots recorded monthly since January 1749 from a NASA website. The data can be read into R using the code below.

```
> S = "http://solarscience.msfc.nasa.gov/greenwch/spot_num.txt"
> sun = read.table(S, header = T)  # sunspots/month data
```

The first line creates a text variable (S) that holds the full name of the website. If you go to this website you'll see that NASA stores these data in four columns, each separated by spaces. We can use the read.table() function and store the dataframe into a variabe (e.g., sun). The following lines of code show how you can plot these data.

```
> plot(sun$SSN, xaxt = "n",type = "l", cex.lab = 1.5,
+       xlab = "Year",
+       ylab = "Number of Spots")  # sunspot number per month
> XA = seq(14,length(sun$SSN)+12,by = 120) # location of
>       # January 1 for multiples of 10 years in this database
> lab = seq(1750,2013,by = 10) # labels for x axis
> axis(1,at = XA, labels = lab, las = 2) # make x axis
```

The challenge in making this graph is that the data are provided in months and years and, since there are thousands of data points, we have to choose what to display on the x-axis. We first tell R not to make tick marks or labels in the plot() function (the argument xaxt = "n" suppressed these). Next, I need to decide where the tick marks and labels will be placed. I've chosen to identify the month of January for each decade, starting with the year 1750. We can make tick marks wherever we want (at = XA) and use the labels (lab) to identify the marks. The las = 2 argument rotates the dates on the x-axis by 90 degrees. These commands are completed using the axis() function. The final product is seen in Figure 2.3. We'll discuss more on how to make graphs in chapter 5.

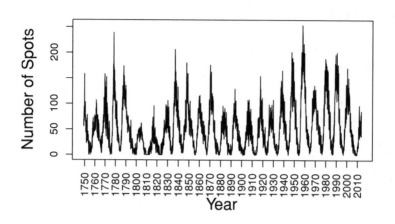

Figure 2.3: The number of sunspots observed between Januar 1749 and August 2013. Data from NASA (`http://solarscience.msfc.nasa.gov/greenwch/spot_num.txt`).

2.4 PROBLEMS

(For questions 1–2). The following values represent the grades earned by six randomly selected students from my biology class.

86.3, 79.9, 92.4, 85.5, 96.2, 68.9

1. Store the grades above in a variable called "`grades`."
2. Determine the arithmetic mean of the grades.

(For questions 3–8). The following are the masses (kg) of 10 raccoons.

2.17, 1.53, 2.02, 1.76, 1.81, 1.55, 2.07, 1.75, 2.05, 1.96

3. Open a new file in Excel and enter "Mass" in cell A1. In cells A2 to A11 enter the data for the raccoons.
4. Save the file as `raccoons.csv` into a working directory for your R projects. You need to choose a different type of file so click on "Save as type" and choose "CSV (Comma delimited) (*.csv)". Remember where you save it! Note, to create `.csv` files in Excel you must specify this file type under the "Save as type" option.

5. Create a script file in RStudio named something like "coons.r" and save this file into the same folder as your `raccoons.csv` file.

6. Set the working directory in RStudio to be the folder where your `raccoons.csv` file is stored (review the steps in Box 1.2 on page 20).

7. Read this data file into an R variable named "my.coons." Once you've read it in check that the name of the variable is correct using the `names()` function. The `names()` function should return the word "Mass" which means that this variable is recognized and contains your data. If you type (`my.coons$Mass`) at at the command line and hit enter you should see your data.

8. Summarize these data using the `summary()` function, being sure to use the `dataframe$header` convention for your data.

9. Find data on the web and import it into R by reading it directly (see section 2.3).

10. Remove all the variables you have created in these exercises. You might choose to include this in your script file but comment it out so that you control when you call this command.

CHAPTER 3

WORKING WITH YOUR DATA

Now that you have data in R we need to think about what to do with them. You should be sure you know where the data came from, what the units are, who collected them, and that they are correct. You also should have some sense of what patterns in the data should look like before you analyze or graph them. This understanding can help you detect whether there are problems with your data.

3.1 ACCURACY AND PRECISION OF OUR DATA

When working with numbers we need to consider the difference between accuracy and precision and understand how they apply to our data. We also need to be careful in how we report our numbers.

1. **Accuracy**. When we use an instrument to measure something, we want our measurement to be as close to the actual value as is possible. The closeness of this measurement to the real value is referred to as "accuracy."

2. **Precision**. This is the similarity of repeated measurements of the same value. If our measurements of the same value, for instance, are quite similar then precision is high. Interestingly, we can have high precision and low accuracy. Alternatively, we could actually have high accuracy (in one measurement) while our precision might be low (we just happened to have been lucky).

3. **Significant digits**. We need to be careful in how we report summary statistics. For instance, if we have the mass of 1000 cats, all recorded

to the tenth of a kilogram (e.g., 2.2 kg), our mean of those cats can't
have an outrageous number of significant digits (e.g., 2.8548 kg). We
often accept one more significant digit than the number with the lowest
number of significant digits when reporting summary statistics (e.g., the
mean). There are many ways to help us with this in R. Below you can
see how the function round() and signif() work.

```
> a = 3.141592654 # pi
> b = 3141592654 # pi x 10^9
> round(a,2) # control number of decimal places displayed

[1] 3.14

> signif(a,2) # this controls significant digits

[1] 3.1

> round(b,2)

[1] 3141592654

> signif(b,2)

[1] 3.1e+09
```

You should allow R to do all the calculations with all the significant digits
it wants to use. However, you should report your data using the correct
number of significant digits for the standards of your particular subdiscipline
(or whoever is evaluating your work).

3.2 COLLECTING DATA INTO DATAFRAMES

In the previous chapter we discussed reading data into RStudio from comma-
separated variable files (usually ending with .csv). When we read in a file and
store it in a named variable, that variable itself points to a structure called a
"dataframe." Dataframes, like Excel spreadsheets, can contain many different
types of data stored in different columns. Each column should have its own,
unique name, preferably have no empty spaces, and contain data that are all
of the same type (e.g., all numbers or all character strings).

If our data are in separate variables they can be difficult to work with. If,
for instance, we had mass data for seeds from 20 different species then we'd
have 20 different variables. A dataframe conveniently can be used to gather
data together into a single variable. This is similar to a "database" or even

an Excel spreadsheet. To do this we send our variables to the `data.frame()` function and store the returned object into a variable. Let's create a dataframe that's made up of two array variables. We'll use the male and female height data from the previous chapter:

```
> males = c(1.72, 1.69, 1.70, 1.66)
> females = c(1.55, 1.62, 1.58, 1.53)
```

We can combine these into a single dataframe in a variety of ways. The easiest way is as follows:

```
> my.dat1 = data.frame(males,females)
> my.dat1

  males females
1  1.72    1.55
2  1.69    1.62
3  1.70    1.58
4  1.66    1.53
```

We can see we now have our height data for males and females stored in a single variable called `my.dat1`. On the left R provides the row numbers. Then we see the male and female data in separate columns. If we want to see just the males we can use the $ symbol like this:

```
> my.dat1$males

[1] 1.72 1.69 1.70 1.66
```

This is how we can get just one of our variables as an array. The last thing we should check out are summary statistics for these datasets. Instead of doing it separately for `males` and `females` we can send the entire dataframe to the `summary()` function. Here's what happens:

```
> summary(my.dat1)

    males           females
 Min.   :1.660   Min.   :1.530
 1st Qu.:1.683   1st Qu.:1.545
 Median :1.695   Median :1.565
 Mean   :1.692   Mean   :1.570
 3rd Qu.:1.705   3rd Qu.:1.590
 Max.   :1.720   Max.   :1.620
```

Let's see how we might build a dataframe that contains different treatments. I use the random seed (`set.seed(100)`) so you can try it and get the same numbers.

```
> set.seed(100) # use 100 and we will have the same numbers
> N = 1:10 # N is an array of 10 numbers (1,2,3,...9,10)
> dat = rnorm(10) # create 10 random nums from
>        # the standard normal distribution
> trmt = rep(c("A","B","C","D","E"), each = 2) # our trmts
```

The `rep()` function above creates an array of repeated values. We create a string of characters (A through E) and make two of each character (`each = 2`). We now send the three arrays (`N`, `trmt`, and `dat`) to the `data.frame()` function and store the result in the variable `my.dat2`.

```
> my.dat2 = data.frame(N,trmt,dat) # creates a dataframe
```

We can look at the beginning, or "head," of the resulting dataframe `my.dat2` with this command:

```
> head(my.dat2) # just the first six lines of my.dat2

  N trmt         dat
1 1    A -0.50219235
2 2    A  0.13153117
3 3    B -0.07891709
4 4    B  0.88678481
5 5    C  0.11697127
6 6    C  0.31863009
```

You should now have a dataframe with 10 rows and three columns (ignoring the row numbers). If you type `my.dat2` at the console and hit <enter> you should see all of your data in the dataframe. In the next four sections we will work to manipulate our data so that we can perform different types of graphing and statistical procedures.

3.3 STACKING DATA

Statistics packages usually have two ways of treating multivariate data. These are called "stacked" and "unstacked." You are probably familiar with "unstacked" data. Data in this format will have a separate column for each variable. If, however, we have more than two columns of data (e.g., 20 columns!)

then dealing with all of these becomes tedious. To avoid this, statistics programs, like R, expect data to be "stacked." Let's see the difference in these formats using data.

```
> males = c(1.72, 1.69, 1.70, 1.66)
> females = c(1.55, 1.62, 1.58, 1.53)
> my.dat = data.frame(males,females) # must be a dataframe
> my.dat # unstacked

  males females
1  1.72    1.55
2  1.69    1.62
3  1.70    1.58
4  1.66    1.53
```

The data can be converted into the stacked format by sending the unstacked data to the **stack()** function. The stacked data are returned as a dataframe.

```
> stacked.dat = stack(my.dat)
> stacked.dat

  values     ind
1   1.72   males
2   1.69   males
3   1.70   males
4   1.66   males
5   1.55 females
6   1.62 females
7   1.58 females
8   1.53 females
```

This is great but the names are ugly ("values" and "ind"?). We can tell R what names we'd really like. We can do that with the following:

```
> names(stacked.dat) = c("height","gender")
```

We can see that the names have changed by simply asking R to display the names:

```
> names(stacked.dat)

[1] "height" "gender"
```

3.4 SUBSETTING DATA

Sometimes we want to pull out a subsample from our data frame. Here's how we can do this using the `subset()` function if we want just the data for males:

```
> subset(stacked.dat, gender == "males")

  height gender
1   1.72  males
2   1.69  males
3   1.70  males
4   1.66  males
```

We simply send the `subset()` function our dataframe (`stacked.dat`) and then tell it what we want (e.g., just "males"). We need to use the double equals sign ("==") which means "equal to." The single equals sign is an assignment (make the left argument take on the value of the right argument). A common example of needing to subset data like this is when we want to test a sample group for normality (see chapter 4). In the above example we can send that result of the heights of males to another function. Another use might be if we want to see just those records with `height` $\geq$ 1.6:

```
> subset(stacked.dat, height >= 1.6)

  height  gender
1   1.72   males
2   1.69   males
3   1.70   males
4   1.66   males
6   1.62 females
```

and discover that we have four males and one female that meet this criterion. The `subset()` function requires we send it some sort of structure, be it an array or, in our case above, a dataframe. It then returns the same sort of structure. What if we want just the height data of those individuals who are at least 1.6 m in height? We can extract columns and rows from dataframes like this:

```
> temp.dat = subset(stacked.dat, height >= 1.6)
> temp.dat$height
```

```
[1] 1.72 1.69 1.70 1.66 1.62
```

We can use these data just like any array of data. For instance, perhaps we want the mean and standard deviation of just the heights of females in the dataframe called `stacked.dat`. Here's how we do that:

```
> F.dat = subset(stacked.dat, gender == "females")$height
> mean(F.dat)
```

```
[1] 1.57
```

```
> sd(F.dat)
```

```
[1] 0.0391578
```

3.5 SAMPLING DATA

When setting up experiments we often need to place individuals randomly into groups. The `sample()` function works well for this. Let's imagine we have a set of individuals (e.g., petri dishes or organisms) that we want to place randomly into treatment groups. For our experiment we might imagine that there are 50 individuals that we want to place randomly into five groups with 10 individuals each. How do we randomly choose 10 individuals for the first treatment group? Here's one way:

1. Number each individual, from 1 to 50.
2. Randomize the list of numbers:

   ```
   > N = 1:50 # an array of integers from 1 to 50
   > S = sample(N)
   ```

S now contains a randomized list of numbers from 1 to 50. Note, if you do this your own numbered list will likely be in a different order. We can place the first 10 individuals in this list into our first treatment group. We can see the IDs to the first 10 individuals like this:

```
> S[1:10] # the ID for the first 10 individuals
```

```
 [1] 27 35 26 36 20  8 34 38 24 12
```

We can list the second group of 10 individuals like this:

```
> S[11:20]
```

```
[1] 46 37 14 47 48 32  7 21 49  5
```

Alternatively, you can create a matrix of your data so that each group is in its own column, like this:

```
> matrix(S,ncol = 5)
```

```
       [,1] [,2] [,3] [,4] [,5]
 [1,]   27   46   10   50   17
 [2,]   35   37   39    4    6
 [3,]   26   14   22   28   41
 [4,]   36   47   23   18   29
 [5,]   20   48   16   30   11
 [6,]    8   32   13   42   43
 [7,]   34    7   19    2   40
 [8,]   38   21   33    3   15
 [9,]   24   49   31   45    1
[10,]   12    5   44   25    9
```

That creates a matrix with five columns. Each column represents a treatment group with 10 individuals each.

3.6 SORTING AN ARRAY OF DATA

The function sort() will sort a list into increasing (default) or decreasing order. We can look at this with our numbers from above. If we needed to see the 10 individuals in the second group from above but wanted them in increasing order for simplicity we could do this:

```
> sort(S[11:20])
```

```
[1]  5  7 14 21 32 37 46 47 48 49
```

Now the numbers are in order and we can grab our labeled petri dishes, test tubes, or organisms in order. We can sort them into decreasing order like this:

```
> sort(S[11:20],decreasing = TRUE)
```

```
[1] 49 48 47 46 37 32 21 14  7  5
```

3.7 ORDERING DATA

The function **order()** returns the *index* for a sorted ordering of objects. Let's use the list of the random numbers from 1:50 (variable S). If we send that random ordering of the numbers through the **order()** function what happens? The function will return another, seemingly unrelated list of what appears to be random numbers. Those numbers are the ordered *indices* for the numbers in increasing order. Perhaps this will help:

```
> S # the randomized list of numbers

 [1] 27 35 26 36 20  8 34 38 24 12 46 37 14 47 48 32  7 21
[19] 49  5 10 39 22 23 16 13 19 33 31 44 50  4 28 18 30 42
[37]  2  3 45 25 17  6 41 29 11 43 40 15  1  9

> order(S) # the indices of the numbers; lowest to highest

 [1] 49 37 38 32 20 42 17  6 50 21 45 10 26 13 48 25 41 34
[19] 27  5 18 23 24  9 40  3  1 33 44 35 29 16 28  7  2  4
[37] 12  8 22 47 43 36 46 30 39 11 14 15 19 31
```

Notice that the first number is 49. That means that the 49th element of S is the first number in the ordered list. In the original data we see that the 49th element is 1.

For another example, assume we have the names of four people that are not sorted alphabetically. Let's store them in a variable called *P*:

```
> P = c("Bob","Sally","Fred","Jane") # 4 people
> P # just to see our array P before sorting

[1] "Bob"   "Sally" "Fred"  "Jane"
```

Note that Bob is first, or $P[1]$, Sally is $P[2]$, and so forth. We can sort them into ascending (alphabetical) order like this:

```
> sort(P)

[1] "Bob"   "Fred"  "Jane"  "Sally"
```

We can, alternatively, leave them in their given order and get a list of their indices within the array P for the alphabetically ordered list.

```
> order(P)
```

```
[1] 1 3 4 2
```

These numbers represent the indices that would create our order list in alphabetical order. For instance, Bob is first in the list so the index "1" identifies him. The ordered list then has the third element in the P array, which is Fred. We can use this ordered list of indices to order our array of names:

```
> P[order(P)] # Note: Fred is 2nd in alphabetized list
```

```
[1] "Bob"    "Fred"   "Jane"   "Sally"
```

3.8 SORTING A DATAFRAME

You might have a large dataframe that you've entered yourself or downloaded from the Internet that you want to sort or change the current sorting order. To do this you need to select one more of the variables to serve as the key or keys on which you will sort the dataframe. Let's create a dataframe and sort it.

```
> trmt = c("A","B","A","B")
> males = c(1.72, 1.69, 1.70, 1.66)
> females = c(1.55, 1.62, 1.58, 1.53)
> my.dat = data.frame(trmt,males,females) # creates a dataframe
> my.dat
```

```
  trmt males females
1    A  1.72    1.55
2    B  1.69    1.62
3    A  1.70    1.58
4    B  1.66    1.53
```

The treatments (trmt) alternate between A and B. I would like to sort these data by treatment so the A treatments come first; then the B's. Here's how I can do this:

```
> my.dat[order(my.dat$trmt),]
```

```
  trmt males females
1   A  1.72   1.55
3   A  1.70   1.58
2   B  1.69   1.62
4   B  1.66   1.53
```

This command is a bit more complicated. We can see I used the `order()` function and sent it the `trmt` column (A,B,A,B). But that's in square brackets and there's a hanging comma. Recall that square brackets indicate indices for arrays. This dataframe, however, has two dimensions ([row, column]). So, this command is asking R to sort the dataframe by the `trmt` column, which returns the correct ordering for our rows and then uses that ordering for the rows. The comma separates rows and columns. I left the columns entry empty which tells R to implement this ordering over all columns.

For this sorting we only printed the output to the console. We might, instead, want to keep the sorted data for later use. We could store them in a new dataframe or replace the original dataframe by copying the result back into the same variable:

```
> my.dat = my.dat[order(my.dat$trmt),] # replaces my.dat
```

3.9 SAVING A DATAFRAME TO A FILE

Once we have worked our data into a form that we like, be it stacked, unstacked, or sorted, we may want to simply save it to disk. Remember that in R we should use script files so we write lines of code that prepare and analyze our data. We might be quite content reading in a data file that is a complete mess and fixing it up so that it's beautiful. We really don't have to do anything else with the original data because, with no additional work, we would then have our data correctly entered in R. Nonetheless, if you want to share these data with others then you might want to give them a beautiful version. Here's how to write the file to disk. Note that R will write the file to the default working directory. If you don't know where the file will be written you can use the `getwd()` function at the command prompt.

```
> write.csv(my.dat,"Plant heights.csv")
```

This function call will take the structure (above, called `my.dat`) and write it to the file named in the quotes into the current working directory.

3.10 PROBLEMS

(For questions 1–5). Here are the top speeds of five cheetahs in km hr^{-1}:

$$102, 107, 109, 101, 112.$$

1. Enter these data into an array called **cheetahs**.
2. Sort the data in *decreasing* order.
3. Pull out the indices for the data in *increasing* order.
4. Use those indices to report the data in *increasing* order.
5. You discover that the speed gun used to measure the cheetahs had only a precision to 10 km hr^{-1}. Use the **signif()** function to properly report the data.

(For questions 6–11). Use the following data of the masses of mussels in grams for the questions that follow.

Low	Medium	High
12	54	87
32	34	78
22	45	59
19	69	82
27	83	64
31	44	73
25	22	77

6. Enter the data into a spreadsheet as you see them in the table.
7. Read the data into R, storing them in a dataframe called "my.dat."
8. "Stack" the data, storing the result in the same variable.
9. Sort the data in the stacked dataframe using the data instead of using the O_2 concentrations. Do the data seem to sort by O_2 levels this way?
10. Determine the mean and standard deviations of each of the treatments individually (**A-C**), using the **subset()** function.
11. Find the treatment levels for which individual mussel masses are less than 30.

12. Your professor wants to create working groups with the students in a laboratory. There are 20 students and the professor wants five groups of four. If students have ID numbers from one to twenty, randomly place them into the five groups. Provide, as output for your professor, a table with five columns and four ID numbers per group.

4

TELL ME ABOUT MY DATA

After we've collected some data we usually want to do something with them. We usually want to know if they tell us something important about the world. We first, however, need to check them out to see if the data make sense, are reasonable, where the middle is, and how messy the data are. We'll first define what we mean by data because they come in many different forms. We will then look at the "distribution" of the data and calculate some basic "statistics," or summary values, about the data.

4.1 WHAT ARE DATA?

Data are usually numbers (a single number is referred to as a datum), but they don't have to be numbers (e.g., flower colors). Data can be *continuous* and take on any value in a range. We often think of height as being a continuous value because, for instance, we can really be any value between a certain range. It's not really practical, however, because we are limited by the precision of our measuring instrument. *Discrete* data have only certain values that they can be. This is true for the numbers of individuals. So count data are usually discrete.

Sometimes data come in discrete categories that can be ranked. We might not have speeds or sizes of individuals but only know the order in which the types occurred. Sometimes data have discrete categories but cannot be ranked at all. These data would simply have what we call *attributes*, referred to as *categorical* data. An example might be the color morphs of individuals. These types of data also are referred to as being *nominal*, because we are simply naming different types without any ordering.

Data can be described by the type of scale they occur on. Some data are found on an *interval* scale. This means that differences between values can be compared. For instance, the difference between 32° F and 37° F is the same as the difference between 52° F and 57° F. These data, however, lack a true zero point. Alternatively, data on a *ratio* scale exhibit this interval property but also have a real zero point. The mass of organisms or the number of leaves on plants are on this scale. Temperature in degrees F do not exhibit this property (50° F is not twice as hot as 25° F, although I once heard a meteorologist say this!). The Kelvin scale, however, does have a zero point so these values are on a ratio scale.

4.2 WHERE'S THE MIDDLE?

There are many estimates of where the middle is in a set of numbers. The most common descriptors are the mean, median, and mode. We should recognize that these measures of where the middle is are simplifications, or "models," of our data. It matters which one we choose because sometimes these measures are quite different from each other.

THE MEAN ($\bar{x}$)

There are actually many different means. We usually consider the mean to be what's more technically called the "arithmetic mean." This is the average you're probably familiar with, which is the sum of all the values in a sample divided by the number of observations in the sample. We can write this formally using the summation sign (Σ) which means that we add up all our values from the first (i = 1) to the last (i = n) number. The array of numbers is called x, and the subscript (i), therefore, is the index to each of the numbers in the array.

$$\bar{x} = \frac{1}{n} \sum_{i=1}^{n} x_i \tag{4.1}$$

We calculate the arithmetic mean in R for an array of numbers like this:

```
> x = c(5,3,6,7,4)
> mean(x)

[1] 5
```

If we have a frequency table of values we might need the "weighted mean." For instance, we might have seven ones, three twos, and five threes in our data. We might represent our data in the usually way by listing them:

$$1,1,1,1,1,1,1,2,2,2,3,3,3,3,3$$

This is a rather short list of numbers and we might be fine with just getting the mean using this list. Alternatively, however, for long lists we can get the weighted mean using the function `weighted.mean()`. We can calculate the weighted mean like this:

```
> vals = c(1,2,3)
> weights = c(7,3,5) # 7 ones, 3 twos, and 5 threes
> weighted.mean(vals,weights)

[1] 1.866667
```

THE MEDIAN

If the distribution of your data is asymmetric (skewed or lopsided, see Figure 4.1) then the center of your distribution might best be described by the median, which is the 50th percentile (a "percentile" represents the value at which that percentage of values fall below this value). If you sort your data from lowest to highest (or highest to lowest) then the median is the middle value. This is really nice if there is an odd number of values. If you have an even number of values then the median usually is calculated as the arithmetic mean of the two values on either side of the middle. See the following examples which use the `median()` function:

```
> median(c(2,3,4,5,8)) # 4 is the middle value

[1] 4

> median(c(2,3,4,8)) # the middle lies between 3 and 4

[1] 3.5
```

Let's consider an example of where we might want to find the middle of a distribution. In the table below are grades, points received per grade, and the number of grades received by a graduating senior. The grades have been converted from letter grades to the often-used 0–4 point scale (e.g., a B– is a 2.7). Such a student's performance is often reported as a grade point average (GPA). That's the arithmetic mean.

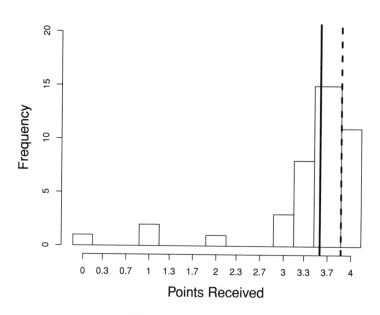

Figure 4.1: The frequency of grades received by a graduating senior. The solid vertical line represents the student's grade point average (GPA) and the vertical dashed line represents the students grade point median (GPM).

Grade	A	A-	B+	B	B-	C+	C	C-	D	F
Points	4	3.7	3.3	3	2.7	2.3	2	1.7	1	0
Number of Grades	11	15	8	3	0	0	1	0	2	1

Now, to calculate this student's GPA, we need to use the `weighted.mean()` function as we did above. What is the student's GPA? This is left as an exercise.

THE MODE

The mode is the most frequent observation (value) in a dataset. This makes sense if the data are discrete (e.g., integers) or are categorical or nominal. If, however, the data are continuous it's likely all values in a dataset are unique. If we plot data in a histogram we usually see some grouping (bin) that has the largest number of observations. In such a circumstance we can think of the mode as either the range of values in this bin or as the center of the bin. This, unfortunately, is dependent on the number of bins we divide our data into. The mode in the GPA data is an A−, and is seen as the most frequent value in the table and in Figure 4.1.

In biology we rarely use the mode as a quantitative measure. Sometimes we think qualitatively about whether data exhibit two modes (e.g., "bimodal distribution") or, perhaps, more modes. If the data are discrete then the mode might be useful to use (the number of leaves on young seedlings plants). If, however, the data are continuous then this is subjective because the size and number of the bins we use affects where and how many modes we could have.

For discrete data we can find the mode using the `table()` function. This function performs a cross-tabulation analysis, which simply gathers up all the different values and counts their occurrences.

```
> set.seed(100) # do this to have the same data
> dat.raw = sample(1:10, 500, replace = T) # get 500 random
> # vals btwn 1 and 10
> dat.table = table(dat.raw) # cross tabulation for the data
> dat.table # here are the data, ordered up by frequency

dat.raw
 1  2  3  4  5  6  7  8  9 10
38 43 52 70 45 49 55 52 40 56
```

The last step is to find the maximum value (in the top row of the output above), which is our mode. To do this we can find which value is the maximum (using the `which.max()` function). We don't want the number of occurrences but, instead, want to know which value has the largest number of occurrences. So we find which of the names in `dat.table` has the largest number of occurrences, using the `names()` function. Finally, we return that result as a number (using the `as.numeric()` function).

```
> as.numeric(names(dat.table)[which.max(dat.table)])

[1] 4
```

If our data are continuous then it's quite possible all our data values are unique so there wouldn't be a mode. We might, however, look at our data using a histogram (the `hist()` function) and see that some range of values occurs most frequently in our dataset. We can actually use the histogram function to do the work for us and get which bin is the most frequent (the tallest bar in the histogram). I show the code below but without presenting the graph.

```
> set.seed(100) # set random seed to get the same answer
> a = hist(rnorm(1000)) # store data from hist() in "a"
> my.mode = a$mids[which(a$counts==max(a$counts))]
> cat("The mode is the bin centered on", my.mode,"\n")
```

```
The mode is the bin centered on 0.25
```

The `hist()` function is quite flexible and allows us to make as many bins as we would like (e.g., use the argument `breaks = 50`, for instance). The line above which identifies the mode using the object data returned by the `hist()` function. We can use this output to find `which()` of the bins is the tallest (has the most observations), using the `max()` function. We're interested in where the midpoint is for this bin and can get that with the `mids` element from the object returned by the `hist()` function. Trust me, you don't want to do that by hand!

Because of this subjectivity in the number of bins made from continuous data we usually avoid using the mode. Instead, we usually use either the mean (symmetric data) and/or the median (symmetric or asymmetric data) to describe the middle of a distribution.

4.3 DISPERSION ABOUT THE MIDDLE

After we've found the middle of our data we next need to understand the spread of our data around this middle. There is a variety of measures used to estimate this spread.

We're going to create a dataset that is normally distributed and then calculate a variety of measures of dispersion. If you use the code I give you then your data will be the same as mine (using the `set.seed()` function). Let's first get 1000 values from the standard normal distribution ($\bar{x} = 0$, $s = 1$).

```
> set.seed(100) # set the random seed
> x = rnorm(1000) # 1000 nums values from the "standard
> # normal distribution"
```

Note that each of the techniques below represents a simplification, or *model*, of our data. This is just like how we looked at different models of the center of a distribution above (e.g., the $\bar{x}$).

RANGE

The range of a dataset is simply the largest value minus the smallest value. It's generally a poor representation of the dispersion since it's sensitive only to the most extreme values, which we might not take much stock in. Sometimes, however, it's just what we need. We'll use this later, for instance, when we want to fit a line to some data but want to do that only over the *range* of the x-variable data (see section 9.2). We can get the range of our x variable like this:

```
> range(x) # returns the smallest and largest values
```

```
[1] -3.320782  3.304151
```

```
> diff(range(x)) # for this, diff() gives us what we want
```

```
[1] 6.624933
```

STANDARD DEVIATION (s)

The standard deviation is a good measure of dispersion about the mean because it has the same units as the mean. It's useful to know what the standard deviation measures. It is approximately the average, absolute difference of each value from the mean (see the section on the variance below). If, for instance, the values are all very close to the mean then s is small. It is easily calculated in R like this:

```
> sd(x) # sd for the standard deviation
```

```
[1] 1.030588
```

VARIANCE (s^2)

The s^2 of these data is actually the square of the standard deviation (s). The sample variance for a group of n numbers is calculated as:

$$s^2 = \frac{\sum\limits_{i=1}^{n}(x_i - \bar{x})^2}{n - 1} \qquad (4.2)$$

We can determine the variance (s^2) in R in either of the following two ways:

```
> sd(x)^2
```

```
[1] 1.062112
```

```
> var(x)
```

```
[1] 1.062112
```

STANDARD ERROR OF THE MEAN (SEM)

Another measure of dispersion that is often used to represent variability is the standard error of the mean (SEM). This measure is actually an estimate of the standard deviation of a sampling distribution, which is the distribution of many means drawn from the same population. This is the foundation of the central limit theorem, which we discuss later (see the central limit theorem on page 197) and is calculated as $SEM = s/\sqrt{n}$, where n is the number of values in our sample. In R we can calculate this as:

```
> SEM = sd(x)/sqrt(length(x))
> SEM
```

```
[1] 0.03259005
```

Some researchers choose to describe the distribution of their data with the SEM. I have actually heard some people like to use this because it's smaller than the standard deviation (note that, by its definition, it will be smaller than sd because $n > 1$). This smaller value might imply that the data have a higher precision. This is not sound thinking. This measure of dispersion should be used if one wishes to show the estimated variability of the standard deviation that would be found if the means of many samples were taken. Note that if we know the number of values in our dataset we can get either s or s^2 from the SEM.

95% CONFIDENCE INTERVALS (95% CI)

Many researchers choose to represent variability as a 95% confidence interval (see section 11.2 for adding these intervals to a barplot). This interval is calculated using the SEM and the t-distribution, as seen below:

```
> n = length(x) # the size of a sample
> CI95 = qt(0.975, df = n - 1)*SEM
> CI95
```

[1] 0.06395281

The qt() function returns a t-value, given that we want a two-sided distribution with N observations. We'll talk more about this in chapter 7.

Using the SEM in this way suggests that we are somehow estimating some measure of dispersion from a large number of samples. The 95% confidence interval is a measure that, if we calculated this from many samples, would capture the true population mean 95% of the time. Erroneously it is sometimes thought of as an interval that says we're 95% confident that the true mean lies in this range. The key to the correct definition lies in its reliance on the SEM.

You might notice that the measures s, s^2, SEM, and the 95% CI rely on $\bar{x}$ (e.g., see equation 4.2 for the variance which relies on $\bar{x}$). This suggests that, for these measures of dispersion to be useful, the $\bar{x}$ must be a good measure of the middle of our data. These measures also value deviations from the $\bar{x}$ (either smaller or larger) equally. Therefore, the utility of these dispersion measures also rely on the assumption that our data are symmetric about the $\bar{x}$. This happens when our data are *normally distributed*. We will learn about statistical tests that rely on the $\bar{x}$, s, and the s^2 and, therefore, find out that they assume the data are normally distributed.

THE INTERQUARTILE RANGE (IQR)

The last measure of dispersion I want to mention is the interquartile range (IQR). This is a range that is based on quartiles, like the median, which is the 2nd quartile or the 50th percentile. The IQR is the 3rd quartile (the 75th percentile) minus the 1st quartile (the 25th percentile). The IQR is a measure of dispersion around the median. Here's how we calculate this in R:

```
> IQR(x) # returns the interquartile range of the x array
```

[1] 1.359288

The calculation of this can be confusing for small datasets so it should be considered an approximation. If the data are continuous and have a large number of values then the measure makes sense.

The IQR is not often used in biology but occasionally we are interested in the boundaries on the middle 50% of observations when the distribution is not necessarily normal. My college uses this range of the SAT scores for entering students by reporting the 25th and 75th percentiles.

Be sure to check with whoever is evaluating your work to find out which measure of dispersion is preferred.

4.4 TESTING FOR NORMALITY

We've already talked about data being "normally distributed." We're now going to discuss this more explicitly and learn how to test whether our data are normally distributed. As we discuss this we should keep in mind that no biological data are truly "normal" in the statistical sense because the mathematical distribution for a normal curve ranges from minus infinity to plus infinity. What that means for heights or masses is, obviously, questionable in real biological systems.

So, technically, we want to know whether the data are not significantly different from a normal distribution before performing a statistical test. The reason for this is that some statistical tests, called "parametric tests," assume that the arithmetic mean of the sample is a good measure of the middle of the data and that the spread of the data is well represented by the standard deviation. If the mean is not in the middle of the distribution and/or the standard deviation poorly represents the spread, then the assumptions of parametric tests will be violated (e.g., the statistical test and, therefore, your conclusions could be terribly wrong!). *Non-parametric* statistical tests do not make the assumption that the distribution of the data is normal. These non-parametric tests, however, do assume that multiple samples, though not normally distributed, have the same distributions (because our null hypothesis is that they come from the same population).

The bottom line is that you must test whether your data are normally distributed before doing a statistical test. Some people think if you're not sure whether the distributions of your samples are normally distributed you should just use non-parametric tests. This is not a safe approach and may lead you to make mistakes in your interpretation of your results.

As we will see, parametric statistical tests include many of the tests with which you are familiar, including regression, correlation, t-test, and analysis of variance tests. Non-parametric tests include the Mann-Whitney U and the chi-square tests.

We may, at times, be able to *transform* our non-normal data into normal data so that we can employ parametric tests. We might do this with growth data, for instance, because many organisms and populations grow exponentially. Log-transforming such data can lead to normally distributed data (see section 4.6).

Our first step to assess a dataset's distribution is to create a graph of the data. Three common graphs for this purpose are histograms, boxplots, and Q-Q plots (see Figure 4.2). All three of these plots are conveniently graphed

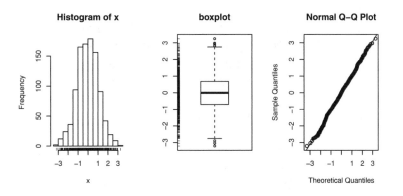

Figure 4.2: Exploratory data analysis visualizations for 1000 values drawn from the standard normal distribution using the function `simple.eda()` from the `UsingR` package. The left graph is a histogram. The center graph is a modified boxplot. The right graph is a Q-Q plot. If the data points in the right graph fall on the straight line then the data adhere to a normal distribution.

for us using the function `simple.eda()`, found in the package `UsingR`. The "`eda`" stands for "exploratory data analysis." Be sure to install that package, if you haven't already, and load it using the `library()` function (see section 1.6).

```
> simple.eda(rnorm(1000))
```

All three visualization approaches can help us evaluate whether our data are normally distributed. I, however, encourage you also to use the Shapiro-Wilk test (the `shapiro.test()` function) to help you decide whether data are normally distributed. The null hypothesis for normality tests is that the data *are* normally distributed. This means that if we get a p-value greater than 0.05 we do not have enough evidence to reject the assumption of normality (see section 6.6 for a complete explanation of p-values).

Let's assume we have four samples and we are interested in determining their distributions. For three distributions we will sample 100 numbers from known populations: a standard normal distribution (`x1`), a uniform distribution (`x2`), and a gamma distribution (`x3`). We will just create the fourth distribution (`x4`) as seen below.

```
> set.seed(10)
> x1 = rnorm(1000) # 1000 values from standard normal dist.
> x2 = runif(1000) # 1000 values from uniform dist.
> x3 = rgamma(1000,2,1) # 1000 values from a gamma distribution
> x4 = c(rep(3,10),1,2,4,5) # leptokurtic distribution
```

Before we test these datasets for normality it's always a good idea to look at their distributions (see Figure 4.3).

```
> par(mfrow = c(2,2)) # create 2 x 2 graphics panel
> hist(x1, main = "Normal Distribution")
> hist(x2, main = "Uniform Distribution")
> hist(x3, main = "Gamma Distribution")
> hist(x4, main = "Leptokurtic Distribution",breaks = 0:5)
> par(mfrow = c(1,1)) # return the graphics window to 1 pane
```

The upper-left panel of Figure 4.3 looks like a bell-shaped curve, indicative of a normal distribution. The other three, however, seem quite different from a normal distribution. Let's evaluate these samples for normality using the Shapiro-Wilk test for normality.

```
> shapiro.test(x1)$p.value # large p -> fail to reject Ho
```

```
[1] 0.25377
```

```
> shapiro.test(x2)$p.value # small p -> reject Ho
```

```
[1] 2.90682e-19
```

```
> shapiro.test(x3)$p.value # small p -> reject Ho
```

```
[1] 2.789856e-25
```

```
> shapiro.test(x4)$p.value # small p -> reject Ho
```

```
[1] 0.001854099
```

The output from the `shapiro.test()` function suggests that only the `x1` dataset is normally distributed ($p = 0.254$) while the results for `x2`, `x3`, and `x4` suggest that these datasets are not normally distributed ($p \leq 0.05$). This agrees with our original interpretation of the histograms in Figure 4.3.

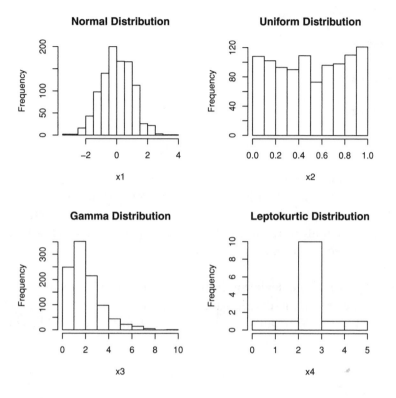

Figure 4.3: Histograms for four distributions for the datasets x1, x2, x3, and x4. We fail to reject the normality null hypothesis only for the distribution in the upper-left panel.

If the data are *not* normally distributed it can be useful to know in what way they violate normality. You've probably heard of a distribution appearing to be "skewed." This is when the data exhibit an asymmetric distribution (see the lower-left panel of Figure 4.3 on page 59). We can test skewness using the skewness() function from the e1071 package. Be sure you've installed the e1071 package (see section 1.6) and loaded it with the library() function. Now you can run the following lines of code to test for skewness.

```
> skewness(x1)
```

```
[1] -0.008931715
```

```
> skewness(x2)
```

[1] -0.03822913

> skewness(x3)

[1] 1.373072

> skewness(x4)

[1] 0

Skewness values close to zero suggest no skew in the data. We see this for the datasets x1, x2, and x4 in Figure 4.3. However, the x3 (lower-left panel of Figure 4.3) is positively skewed (skewed to the right, in the direction the tail is pointing). The uniformly distributed data (upper-right panel of Figure 4.3) are not skewed so are probably not normally distributed for a reason other than skewness.

A last measure of the deviation from normality we will look at is "kurtosis," which is a measure of the distribution's shape away from the standard bell-shaped distribution. You can try this for all four distributions like this:

> kurtosis(x1)

[1] -0.1145157

> kurtosis(x2)

[1] -1.296026

> kurtosis(x3)

[1] 2.356125

> kurtosis(x4)

[1] 1.104286

If kurtosis is near zero then the distribution is "mesokurtic," or is consistent with a bell-shaped curve without a really sharp peak or a flat-topped peak. If the value is a relatively large, positive number then the data are more pointy in the middle than a regular, bell-shaped distribution (x3 and x4). Such a distribution is referred to as being "leptokurtic." If the value is negative then the distribution is relatively flat-topped. As we can see here the kurtosis value for **x2** is more negative than the **x1** sample. The **x2** distribution is referred to as being "platykurtic."

To summarize, we can look at these four statistics, called the first four "moments about the mean," and compare them for these four distributions:

Statistic	x_1	x_2	x_3	x_4
$\bar{x}$	0.011	0.508	1.975	3
s^2	0.984	0.09	1.946	0.769
Skew	-0.009	-0.038	1.373	0
Kurtosis	-0.115	-1.296	2.356	1.104

4.5 OUTLIERS

We all know that outliers are unusual values in a dataset. It turns out they're kind of hard to define but we'll do so formally later in section 5.3 on boxplots on page 70. Outliers are rare (they should occur $< 1\%$ of values drawn from a normal distribution), but are expected to occur if our sample size is large enough. I've heard some people say that outliers should simply be thrown out. This is wrong. You should never discard data simply because the values seem unusual; we need a better reason than that! If a value seems to be totally bonkers then you should investigate why.

Here are some possible explanations for why a data point might appear to be an outlier or otherwise questionable:

1. equipment malfunction,
2. recording error,
3. data entry mistake,
4. general carelessness,
5. extreme environmental conditions at the time of the data collection, and/or
6. a large sample size will yield statistical outliers (they're normal!).

If, after careful consideration, you decide that a value is an outlier that either needs to be removed from your dataset or needs to be changed (e.g., it should be 1.53 not 153) then here's what you should do. If the data were read into R from a file then you should fix the data file and read it back in, if possible. If this is not possible then you need to find it using R and fix it. Let's imagine you have a dataset of 100 values and they're all $1.0 \leq x \leq 2.0$

and one of the values is 153 (it's supposed to be 1.53). You can find its index value in the array (where in the list of 100 numbers it is found). First we need a list of 100 numbers in our range:

```
> x = runif(100)+1 # creates 100 random values (1 <= x <= 2)
```

Next, let's change the 36th value to 153:

```
> x[36] = 153 # makes the 36th value erroneously 153
```

This assigns 153 to the 36th array element. Finally, let's imagine we have our 100 values and know they shouldn't contain a value outside the range $1 \leq x \leq 2$. Here's how we can find the index of any value in an array greater than 2:

```
> which(x > 2) # find index of number(s) that are > 2
```

```
[1] 36
```

R tells us that, in this case, the outlier is the 36th entry in the x array. We can inspect that value like this:

```
> x[36]
```

```
[1] 153
```

and see that the 36th element is out of range. We double-check our lab notebook and see that it's supposed to be 1.53. Here's how you can fix it:

```
> x[36] = 1.53
```

What if the number was recorded correctly but it seems our instrument gave us a faulty value? Then we might want to simply remove this value. If you want to remove the value from the working copy of data in R you can do the following, leaving you with just 99 data points:

```
> x = x[-36] # -36 means removes the 36th value, then copies
>            # the remaining values back into x
```

Remember, if the data are in a spreadsheet you should fix the spreadsheet and read the data back into R, if possible.

4.6 DEALING WITH NON-NORMAL DATA

If the data are not normally distributed what can you do? Before jumping to a non-parametric test, you should investigate whether you can "transform" the data so they become normally distributed. This can be a good idea because if the data are normal upon transformation a parametric statistical test can and should be used. The exponentially distributed data in the histogram in the left of Figure 4.4 are shown without transformation and are then log-transformed, using the natural logarithm with the base e (in R: $log.x1 = log(x1)$, right histogram). Many growth processes in biology lead to exponential distributions which are normalized with this log-transformation.

```
> set.seed(10)
> x1 = exp(rnorm(100, mean = 5, sd = 0.5))
> log.x1 = log(x1) # the natural logarithm of the x1 data
> par(mfrow = c(1,2))
> hist(x1, xlab = "x1 data", main = "",ylim = c(0,30))
> hist(log.x1, xlab = "log(x1) data", main = "",
+       xlim = c(3.5,6.5),ylim = c(0,30))
```

Below is the Shapiro-Wilk test for normality for these two datasets. Recall that the null hypothesis for this test is that the data are normally distributed. Therefore, if $p > 0.05$ then the data appear to be normally distributed (we don't have enough information to reject the H_0).

```
> shapiro.test(x1) # these are NOT normally distributed

        Shapiro-Wilk normality test

data:  x1
W = 0.9076, p-value = 3.286e-06

> shapiro.test(log.x1) # transformed -> normally distributed

        Shapiro-Wilk normality test

data:  log.x1
W = 0.9891, p-value = 0.5911
```

Sometimes you have data with perhaps two or more groups and discover that one sample is normally distributed while the other is not. You can try

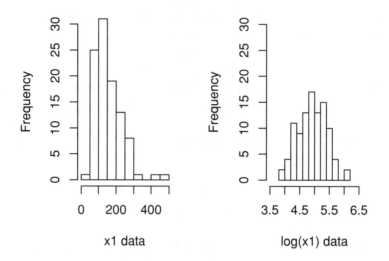

Figure 4.4: Two histograms of 100 data points. The left histogram shows the original data, which are skewed to the right. The histogram on the right shows the data after log-transforming and appear to be normally distributed. The transformation was done with `log(x1)` which, in R, takes the natural logarithm of the `x1` data.

to transform the data and the samples may be normally distributed. What might happen, however, is that the non-normal data become normally distributed and the previously normal data become non-normally distributed. What should you do? This is tricky so discuss this with a statistician or adviser. What can become critically important in such cases is knowing what the underlying distribution should be or knowing how previously collected data were distributed. If the data are usually normally distributed, or should be for a very good reason, then you might assume normality.

4.7 PROBLEMS

(For questions 1–3). The following values represent average lengths of flagella in *Chlamydomonas* (a green alga) as a function of the number of flagella per cell (Marshall et al., 2005). Store these in a variable called `flag`.

$$11.1, 9.9, 8.7, 8.6, 7.0, 4.3$$

1. Calculate the mean, standard deviation, median, and the standard error of the mean (SEM) for these data.
2. Create a histogram, boxplot, and Q-Q plot with a line using the function `simple.eda()` from the `UsingR` package.
3. Test the data for normality using the Shapiro-Wilk test. Does this statistical test seem to support what the visualizations suggest? What concerns do you have about this analysis?

(For questions 4–5). Create a set of 1000 normally distributed random numbers with a mean equal to your height in inches and $sd = 5$. Store these in a variable called `my.dat`.

4. Complete the calculations and graphs from questions 1 - 3 on these data.
5. Get the values for the outliers.

(For questions 6–8). Go to `http://data.worldbank.org/indicator/SP.POP.TOTL` and download the Excel spreadsheet data for populations by country.

6. Save the data to your computer, convert the data to `.csv` file format, and read the data for just 2012 for all countries and store those in an array called `pop`.
7. View these data with the `simple.eda()` function. What does this tell you?
8. Collect the totals for all years and create a plot of how the Earth's estimated population has changed over time.

9. Can you think of three different examples of biological data that are:
    a. continuous?
    b. discrete?
    c. categorical?
10. Can you think of an example other than discussed above where you'd need the range of the data?
11. The following biology grades were received by a student majoring in biology.

Grade Received	A-	A	B-	A	B	B+	C+
Credit Hours	3	3	4	3	1	4	3

What is this student's GPA, based on these grades, accounting for the different weights (credit hours)? Note that letter grades correspond to quality points as follows: A = 4.0, A– = 3.7, B+ = 3.3, B = 3.0, etc.

CHAPTER 5

VISUALIZING YOUR DATA

This chapter is aimed to help you visualize your data. Providing clear, concise graphs of relationships efficiently conveys information to your readers. R, fortunately, is extremely powerful and flexible in allowing us to create a variety of visualizations, from simple scatterplots to full-blown animations and 3-D clickable diagrams. With its basic drawing palette you're actually able to create any visualization you are likely to need. In this chapter we will develop a variety of graphs, enhance those graphs, and discuss the reasons why we might use each of these graph types.

As we begin you will undoubtedly think R is rather primitive. Writing a line of code to generate a graph might even seem crazy! But you will likely soon realize that it is easy and intuitive to use typed words to tell the computer what you want. Yes, it might seem hard at first, but I hope you enjoy the skills you develop to make publication-quality graphs. To see some of what's possible using R check out an R gallery online (e.g., `http://addictedtor.free.fr/graphiques/allgraph.php`http://addictedtor.free.fr/graphiques/allgraph.php).

In this chapter you'll see a variety of basic graphing procedures. The more you play with the different graphs the better you'll get. So, be sure to give them a try.

5.1 OVERVIEW

Before we get started let's create some random data. The following lines of code will create three variables (x, y, and z) that each have 100 values drawn from normal distributions with standard deviations equal to 1. The first line sets the random number seed so that your data will be the same as my data.

This way your graphs should look exactly like those I've created.

Box 5.1. **Optional arguments for visualizations.** The following options, or arguments, can be included to improve the looks of graphs. When provided they must be separated with commas. An example might look like this: > `plot(x,y,xlim = c(0,10))`.

- `xlim = c(low,high)` and `ylim = c(low,high)`. These control the range of the `x-y` axes, respectively.
- `type = "n"`. This produces no points or lines. This is used when you want a little more control by, for instance, adding several sets of data using the functions `points()` or `lines()`.
- `type = "p"`. This is the default for the `plot()` function. If you omit this argument then you'll get open point characters.
- `type = "l"` (that's the letter "el").
- `type = "b"`. This produces both a line and the point symbols.
- `pch = number`. This controls the type of points produced in the `plot()` function. Some of the options for this argument include:

  - `pch = 1`. This is the default and produces a small open circle.
  - `pch = 16`. This produces a small solid circle.
  - `pch = "symbol"`. You can replace "symbol" with a letter, such as "." (a period), or `type = "+"`.

- `lty = number`. This produces different line types. The default (1) is a solid line. A dashed line is 2.
- `lwd = number`. This is the line width. The default width is 1.
- `cex = number`. This number controls the size of the symbol, 1 being the default.
- `cex.lab = number`. This number controls the size of the `x`- and `y`-axis labels. The default is 1 but I usually like 1.5.
- `main = "Your title"`. If you want a title.
- `xlab = "Your x-label"`. Self explanatory.
- `ylab = "Your y-label"`. Ditto.

Additionally, you might consider placing text in your graph using the `text()` function, which might look like this: > `text(x,y,"text")`, where the `x` and `y` values are coordinates in the graph where the text will be centered.

```
> set.seed(100)
> x = rnorm(100, mean = 5) # 100 random nums, mean = 5, sd = 1
> y = rnorm(100, mean = 6) # 100 random nums, mean = 6, sd = 1
> z = rnorm(100, mean = 8) # 100 random nums, mean = 8, sd = 1
```

As you proceed through this section you'll see a number of different types of graphs. Graphs in R are made by sending data variables to graphing functions. We refer to these data, and anything else we send to a function, as "arguments." In the examples that follow you will see how data are sent to the graphing functions and a variety of optional arguments that improve the looks of the graphs. Keep your eyes open for when I do this in the examples. Some of the important arguments used to improve the looks of graphs are summarized in Box 5.1. You can see how this can help improve the scatterplot in Figure 5.1.

```
> par(mfrow = c(1,2)) # make graph window 1 row and 2 columns
> my.dat = c(5,3,6,2,1) # here are my data
> plot(my.dat) # left plot with no graphics parameters
> plot(my.dat, xlim = c(0,10), ylim = c(0,10),
+     xlab = "My x-axis label", ylab = "My y-axis label",
+     main = "My Custom Title", pch = 16, cex = 1.2,
+     cex.lab = 1.5) # a professional-looking graph!
```

In addition to these graphical parameters you might be interested in adding a legend to your graph. This is done with the **legend()** function. Since this is a function it is called separately; on its own line. This can be a little tricky and so you might just steal some code from an online source and tweak it for your purpose. There's an example of the implementation of this on page 76 in section 5.6. You can also read about it and see some examples through the built-in help facility (> **?legend**).

5.2 HISTOGRAMS

Histograms are standard graphs for visualizing the *distribution* of a dataset. This is the best plot for getting a first look at your data. It's quite easy, for instance, to see if the data are *normally distributed* (see Figure 5.2), which is important to know as you begin testing your hypotheses.

```
> par(mfrow = c(1,2))
> hist(x, cex.lab = 1.5, main = "")
> abline(v = mean(x),lwd = 5)
```

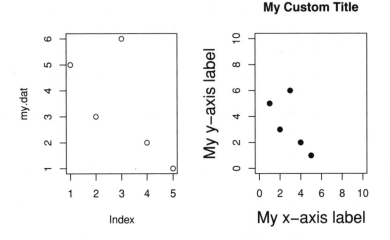

Figure 5.1: An example of a default scatterplot (left) and one with modifications (right). For the graph on the right I've added arguments that make it more professional looking (see Box 5.1).

```
> hist(x, cex.lab = 1.5, main = "", breaks = 34,ylim = c(0,10))
> abline(v = mean(x),lwd = 5)
> par(mfrow = c(1,1))
```

You can control the number of bins that your data are placed into with the "breaks" argument (e.g., breaks = 25). You might first just examine the default, which is usually fine. If the data are normally distributed they should form a bell-shaped distribution.

5.3 BOXPLOTS

Boxplots, like histograms, can be used to show the distributions of data. They sometimes are a preferred graph over barplots because they show the relationship among samples and show data ranges, distributions, quartiles, and the median.

Boxplots generally show a box with a line through the middle. The top of the box is the 75th percentile, the middle line shows the median, which also is the 50th percentile, and the bottom of the box is the 25th percentile. The range between the 75th and 25th percentiles is referred to as the "interquartile

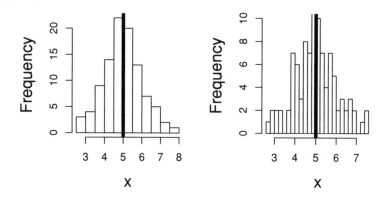

Figure 5.2: Two histograms of the x data. On the left I let R select the number of bars to graph. The thick, vertical line is placed on the graphs at the mean using the `abline()` function. For the histogram on the right I requested 34 bars (`breaks = 34`).

range" (IQR). A percentile is a value at which that percentage of observations is below. Therefore, the 50th percentile is the value at which 50% of the observations are *below*. Interestingly, a perfect score of 800 on an SAT test usually represents *only* the 95th percentile because 5% of those who take the exam get perfect scores. The whiskers may extend above the 75th percentile to the largest value but must not exceed 1.5 times the interquartile range (IQR) above this 75th percentile. There also can be a whisker below the 25th percentile in the same manner. Data points that lie further than the whiskers are called "outliers." So, we now have a formal definition of an outlier:

What is an outlier?

An outlier is a value that is more extreme than 1.5 times the interquartile range above or below the 75th or 25th percentiles, respectively. Or, more simply, they are the points beyond the whiskers of a boxplot when graphed using R.

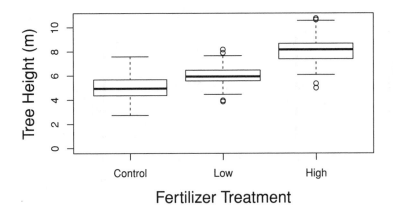

Figure 5.3: A boxplot of tree growth under three different conditions (A, B, and C). Note that the y-axis extends down to zero (`ylim = c(0, ...)`).

In Figure 5.3 I've plotted x, y, and z, pretending they are the heights for three groups of trees grown under different conditions (treatments). Treatment levels "B" and "C" both have outliers.

```
> boxplot(x,y,z,names = c("Control","Low","High"),
+         xlab = "Fertilizer Treatment",
+         ylab = "Tree Height (m)",
+         cex.lab = 1.5,ylim = c(0,max(c(x,y,z))))
```

5.4 BARPLOTS

Barplots generally show a summary value for continuous variables. The heights of the bars almost always represent the arithmetic means. Sometimes bars are displayed horizontally but this is usually not best because the response variable (the variable that was measured) should be shown on the y-axis. The function `barplot()` needs just the heights of each bar in a single array. In the example, I use the means, placed into a new variable called `Ht` (see Figure 5.4). Note that barplots should extend down to zero on the response variable axis (usually the y-axis).

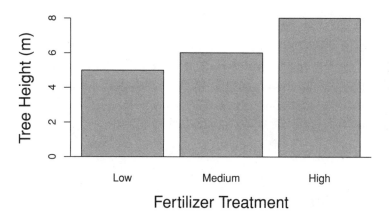

Figure 5.4: A barplot showing the means of individuals of a tree species grown under three different levels of fertilizer. These are the same data as in Figure 5.3. Does something seem to be missing?

```
> Ht = c(mean(x),mean(y),mean(z))
> barplot(Ht,
+         xlab = "Fertilizer Treatment",
+         ylab = "Tree Height (m)",
+ names = c("Low","Medium","High"), cex.lab = 1.5)
> abline(h=0)
```

You need to be careful when using a barplot. You've seen these often but they greatly simplify your data, reducing all the data in a sample to a single value. If you use barplots you should consider including "error bars" which represent the variability of the data (discussed at length in section 11.2 on page 164). Without error bars we can really miss a lot of information, compared to what boxplots provide.

Let's compare the differences between a plain barplot and a boxplot for two samples of data with the same mean and different levels of variability (see Figure 5.5). The datasets are the same for both graphs. You can see that the barplot on the left showing only the mean suggests there is no difference between the two samples while the boxplot on the right shows that the two datasets are really quite different.

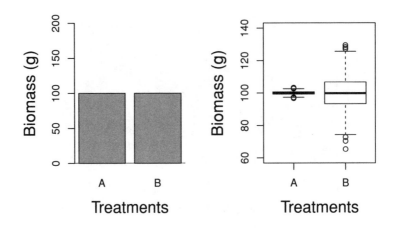

Figure 5.5: On the left are two barplots of the mean biomass for two samples of plants (treatments A and B). These treatments appear to have resulted in no differences. The boxplots on the right show the same data but reveal that the samples are really quite different from each other when we take into account the variability of each sample. Treatment samples A and B are the same in both plots.

```
> a = rnorm(1000,mean = 100,sd = 1)
> b = rnorm(1000, mean = 100, sd = 10)
> par(mfrow = c(1,2))
> barplot(c(mean(a),mean(b)),ylim = c(0,200),
+   xlab = "Treatments", ylab = "Biomass (g)",
+   names = c("A","B"),cex.lab = 1.5)
> abline(h=0)
> boxplot(a,b, ylab = "Biomass (g)",names = c("A","B"),
+   ylim = c(60, 140), cex.lab = 1.5, xlab = "Treatments")
> par(mfrow = c(1,1))
```

Sometimes we want to create a barplot that compares observations over two factors. This is often needed when we have data we are analyzing using a two-way analysis of variance (see section 8.3 on page 123). If we want to do that then we need to decide how to gather up our data. Let's imagine we have the following average Medical College Admission Test (MCAT) scores for males and females for biology and non-biology majors.

Gender	Biology	Non-Biology
Males	36.2	28.9
Females	36.3	29.5

If we wish to plot these means using a barplot we have to decide which factor goes on the x-axis and which factor is represented by the "trace" factor (that's the factor that's paired and shown in the legend). The factors can be interchanged and you'll have to decide which one makes the most sense for your data. For these data I have done it both ways (see Figure 5.6). Here are the data and how they are combined into a matrix for the `barplot()` function.

```
> males = c(36.2,28.9) # first is biology, then non-bio majors
> females = c(36.3,29.5)
> M = matrix(c(males,females),byrow = T, nrow = 2)
```

To create the side-by-side barplots we need the data to be in a matrix. The command above stores the data as a matrix in the variable M. The default for the `matrix()` function is to combine the data by columns. I like to enter the data in rows and so added the argument `byrow = T`.

```
> par(mfrow = c(1,2)) # graphics panel with 1 row, 2 columns
> barplot(M,beside = T, names = c("Biology","Non-Biology"),
+         ylim = c(0,60), xlab = "Major", cex.lab = 1.5,
+         ylab = "Mean MCAT Score",
+         legend.text = c("Males","Females"))
> abline(h=0)
> # Alternatively,
> bio = c(36.2,36.3) # first males, then females
> non.bio = c(28.9,29.5)
> M = matrix(c(bio,non.bio),byrow = T, nrow = 2)
> barplot(M,beside = T, names = c("Males","Females"),
+         ylim = c(0,60), xlab = "Gender",
+         cex.lab = 1.5, ylab = "Mean MCAT Score",
+         legend.text = c("Biology","Non-Biology"))
> abline(h=0)
```

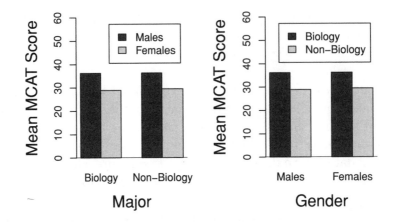

Figure 5.6: Two barplots that represent the same data in two different ways. The graph on the left shows us mean MCAT scores for biology versus non-biology majors, separated by gender. The graph on the right shows us the same data for males and females, each separated by major. (Note that the scores were conceived by me for my biology audience.)

5.5 SCATTERPLOTS

Scatterplots are used to show the relationship between two continuous variables. You might just show points for viewing relationships or you might add a best-fit line to a scatterplot (see section 9.2 on linear regression) when there's a statistically significant dependency that you want to model. Such plots are made with the `plot()` function. I have made a simple scatterplot showing the relationship between two sets of random data drawn from a uniform distribution (Figure 5.7, left panel). The data in the right panel were made using a function ($y = 3x + 2$) along with a little random noise added (uniform random deviate `delta`: $-5 \le \delta \le 5$).

```
> par(mfrow = c(1,2))
> set.seed(2)
> x = runif(10)
> y = runif(10)
> plot(x,y, pch = 16, xlab = "X Variable",
```

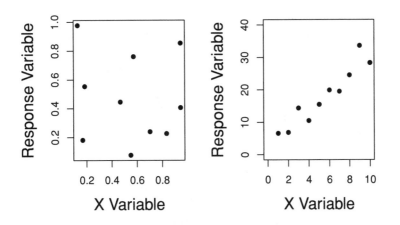

Figure 5.7: The left panel shows a scatterplot that appears to lack a relationship between two groups of data drawn from uniform random distributions. On the right is a scatterplot for data that follow the relationship $y = 3x + 2$.

```
+    ylab = "Response Variable", cex.lab = 1.5)
> # Here are some data that exhibit a relationship
> x = 1:10
> delta = (runif(10) - 0.5) * 10 # random noise
> y = 3*x + 2 + delta
> plot(x,y, pch = 16, xlab = "X Variable", xlim = c(0,10),
+       ylim = c(0,40), ylab = "Response Variable",
+       cex.lab = 1.5)
```

5.6 BUMP CHARTS (BEFORE AND AFTER LINE PLOTS)

Sometimes we have data with measurements on subjects before and after applying a treatment. We might, for instance, measure reaction time to a stimulus in a variety of organisms. We then train them and try the treatment to the organisms again to see if they changed their response rates. We can make a nice graph in R by following these steps:

1. Enter the data into a spreadsheet in Excel. It should have two columns.

2. Read the data into R. It comes in as a dataframe.
3. Create an empty plot.
4. Draw lines that connect the paired data (before and after).
5. Draw points at the ends of each of the lines, if we want.

An example of such a graph is shown in Figure 5.8 on page 79.

```
> N = 20 # number of individuals in sample
> time1 = rnorm(N, mean = 3)
> time2 = rnorm(N, mean = 7)

> plot(0, xlim = c(0.75,2.25),ylim = c(0,10),
+       type = "n",xaxt = "n", xlab = "Time",
+       ylab = "Measurement", cex.lab = 1.5)
> axis(1,at = c(1,2),labels = c("Before","After"),
+       cex.axis = 1.25)
> for (i in 1:length(time1)) { # add the lines one at a time
+   lines(c(1,2),c(time1[i],time2[i]),lty = 1)
+ }
> points(rep(1,length(time1)),time1, pch = 16, cex = 1.5)
> points(rep(2,length(time2)),time2, pch = 17, cex = 1.5)
```

I'm now going to make an empty plot (`plot(0,...)`) using "`type = "n"`"
and then use a "`for`" loop to fill in each line one at a time. This looping is
complicated and is explained in more detail in chapter 12 on page 193. The
resulting graph is Figure 5.8. A good time to use this type of graph is seen
later in section 7.2 on page 106.

5.7 PIE CHARTS

Pie charts are a type of graph that can be used when data are categorical and,
usually, sum to 100%. However, simply stated, pie charts are rarely used in
biology. They are challenging because the amounts in each slice are too hard
to compare. To overcome this problem pie charts often display the numerical
value of the size of the slice, as well. Showing both the graph and the data
values is redundant and should be avoided. If there are few slices in the pie
then those data might best be presented in a table.

```
> x1 = c(10,20,30,40)
> x2 = c(5,25,35,35)
```

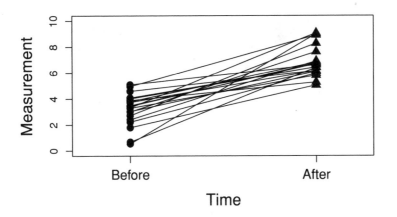

Figure 5.8: A "bump chart" showing lines that connect observations made on individuals before and after some treatment. We provide a graph like this so that a reader might see an overall trend between subjects. Notice that each *line* represents a measurement from a *single* experimental unit so these data can't be analyzed as though the before and after points represent independent samples (see section 7.2 on page 106).

```
> my.dat = matrix(c(x1,x2),nrow = 2,byrow = T)
> my.col = gray(seq(0,1.0,length=4))
> layout(matrix(c(1,2,3,3), 2, 2, byrow = TRUE)) # set layout
> pie(my.dat[1,], col = my.col,radius = 1,
+     main = "Pie Chart 1")
> pie(my.dat[2,], col = my.col,radius = 1,
+     main = "Pie Chart 2")
> leg.txt = c("Pie 1 data","Pie 2 data")
> barplot(my.dat,beside = T, ylim = c(0,50), names = 1:4,
+        col = gray.colors(2), ylab = "Height (cm)",
+        xlab = "Groups", cex.lab = 1.5, main = "Barplot")
> legend("topleft",leg.txt, fill = gray.colors(2))
> abline(h=0)
```

The two pie charts show different data but it is difficult to distinguish the differences between these two datasets. The function `pie()` takes as an argu-

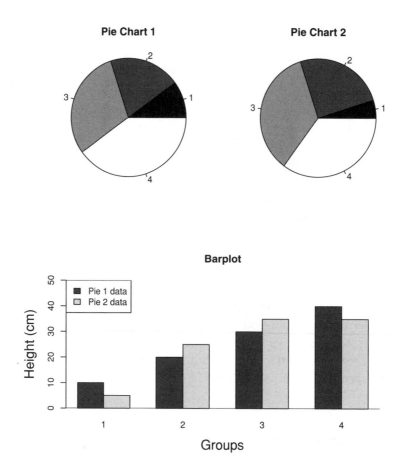

Figure 5.9: The upper panel shows two pie charts with slightly different data. The differences between each group across the two pie charts are difficult to see. You never want to make your reader work to understand your results. These data are presented more clearly in a barplot below the two pie charts. The two datasets are graphed using a side-by-side alignment (`beside = T`) for each group, allowing the reader to make comparisons beteen the two datasets.

ment the radius of the chart so you can have pie charts that proportionately represent different data. Usually, however, the data are better represented using another graphing procedure, such as a barplot (Figure 5.9), which more clearly shows differences between groups.

5.8 MULTIPLE GRAPHS (USING PAR AND PAIRS)

You have already seen several examples where I've managed to place multiple graphs into a single graphics panel (e.g., the pie graphs in Figure 5.9). You can even mix graph types. This is done using the **par()** function, which controls the making of graphs. You can stack graphs, such as histograms, on top of each other (as seen in Figure 5.10) with the **par()** function using a matrix approach to set up the graphics window. In Figure 5.10 I've created a graphics window with 3 rows and 1 column (> **par(mfrow = c(3,1))**. If you're trying to help your reader see similarities or differences in your data you should show them on the same scale (e.g., use the > **xlim = c(xmin,xmax)** option in your plot command). I actually extend these ranges a little more with the **floor()** and **ceiling()** functions (see code). In this example I used the same limits so the range on the **x-axes** are the same so only the data change among the graphs.

```
> x1 = rnorm(20, mean = 7)
> x2 = rnorm(20, mean = 8)
> x3 = rnorm(20, mean = 9)
> # min() gets smallest value; floor() rounds down
> xmin = floor(min(c(x1,x2,x3)))
> # max gets highest value; ceiling() rounds up
> xmax = ceiling(max(c(x1,x2,x3)))
> par(mfrow = c(3,1)) # set graphic window: 3 rows, 1 col
> hist(x1,xlim = c(xmin,xmax))
> hist(x2,xlim = c(xmin,xmax))
> hist(x3,xlim = c(xmin,xmax))
> par(mfrow = c(1,1)) # reset the graphics window
```

Another useful and related graph is a matrix plot which shows the relationships among several variables simultaneously (Figure 5.11). You can make these using the **pairs()** function. This is actually really handy to use because it allows you to look for patterns among continuous data and *not* actually do a hypothesis test. This is a type of "data mining" without actually testing hypotheses.

```
> x = data.frame(x1,x2,x3)
> pairs(x)
```

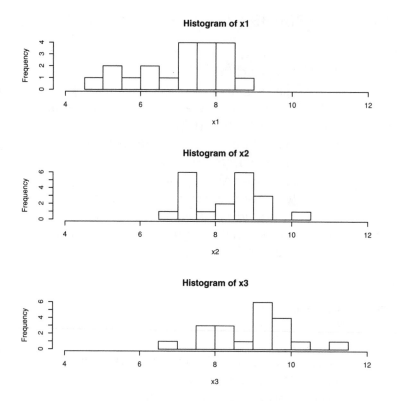

Figure 5.10: A set of three histograms, stacked one above the other, with the help of the `par(mfrow = c(3,1))` command (three rows and one column).

5.9 PROBLEMS

(For questions 1–2). Use the built-in dataset called "women" which contains height and weight data for women in the USA (1975).

1. Look at the dataset for women in the console (> `women`). Also, check out the help page on the `women` dataset (> `?women`). Create the most appropriate visualization for these data.
2. Explain why, biologically, the data appear to show the relationship they do between weight and height.

(For questions 3–4). The built-in dataset "precip" contains annual rainfall data for cities in the USA. Use it to answer the following problems.

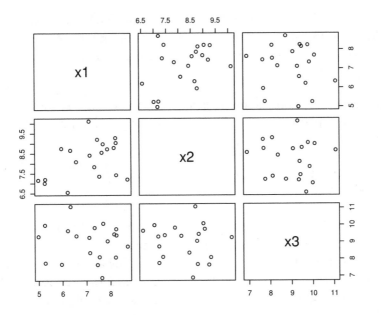

Figure 5.11: A matrix scatterplot using the `pairs()` function. The main diagonal shows no data but instead identifies the three variables (x1,x2, and x3). In the upper-left corner you see x1. The two graphs of data to the right of the x1 box are graphing x1 on the y-axis. The two plots directly below the x1 box have x1 on the x-axis. As an example, the graph in the middle of the bottom row shows x2 on the x-axis and x3 on the y-axis. The data are mirrored on either side of the main diagonal. It's up to you to consider which makes the most sense to consider.

3. How many cities are represented in the dataset? Determine this through the help feature for this dataset and by determining the `length()` of the dataset.

4. Examine visually the distribution of the rainfall amounts in these cities. Use the `simple.eda()` function from the `UsingR` package.

(For questions 5–8). Create a set of 100 normally distributed random numbers with a mean of 10 and a standard deviation of 3. Call these data `my.rand.dat1`.

5. Create a histogram of these data.

6. Create a boxplot of these data.

7. Create a side-by-side graph of a histogram (left) and boxplot (right).

8. Create a second set of random data, called `my.rand.dat2`, and make a boxplot with both samples in a single graph (like the right panel in Figure 5.5).

9. Below are data on the number of bacterial colonies found on two sets of treatment plates; one set of six plates was grown on control median, the other on enhanced growth media.

```
control = 2,3,4,5,6,7
trmt = 5,3,4,5,6,9
```

Provide a side-by-side boxplot of these data. Add labels for the axes, individual treatment levels, and add a title.

10. A student conducts an analysis of squirrel behavior on her campus. She discovers that they spend 10% of their time foraging, 30% of their time grooming, 40% of their time playing, and 20% of their time sleeping. Create a pie chart of these data and a side-by-side barplot for comparison.

CHAPTER 6

THE INTERPRETATION OF HYPOTHESIS TESTS

The process of doing science is quite old. It is a way to understand our universe. It is not the only way. Perhaps you've had the unpleasant experience of running a bit late and discovering your car won't start. You might jump out of the car, pick up sand, throw it toward the east, and say a prayer. Then jump back in the car and give it another go. It might start!

Undoubtedly this approach of doing things and having a good outcome has happened before. I can think of several examples where a behavior might have caught on because of its timing prior to a desired outcome, such as rain dances to bring on rains, crossing fingers for "good luck," and baseball hitters going through their routines before batting. These behaviors might help produce the desired outcome but, when good results follow, they might better be explained as simple coincidences. What separates the above methods from science is that science focuses on using empirical evidence, acquired through rigorous observations and/or experiments to illuminate the mechanisms that govern how natural systems work. In this chapter we'll briefly explore the process of getting scientific information that we might use in our tests of hypotheses.

6.1 WHAT DO WE MEAN BY "STATISTICS"?

This book is purported to be about statistics and visualizations. A visualization is simply a representation of information, often of data, usually presented in the form of a graph (see chapter 5). Understanding what we mean by "statistics," however, is more complicated. I define "statistics" as the ***process*** *of gathering, summarizing, and interpreting qualitative and quantitative information*. In biology our process of getting information usually follows the standards of the scientific method and relies on proper experimental design (see section 6.4). Numbers by themselves do not constitute statistics, and certainly not if there is no context. When we attach information to numbers we can have something with a great deal of meaning.

The simplest statistics are summary values, which are called "descriptive statistics." You encounter these all the time, such as the average score on an exam, the fastest time in a race, or even the approval rating of the president's job performance. I am watching a sporting event right now as I write. The announcer said that the team was relatively old with an average age of 25. What does that mean? They all might be 25 or none be 25. They might have the youngest and/or the oldest athletes in history on the team! The announcer has provided a model (or simplification) of the ages of the athletes on the team. We, however, have been given very little information. Numbers can describe something quantitatively but they need our help to make them meaningful.

In biology we are often interested in knowing something about a *population*, which usually includes everything in that group (e.g., all wolves on Earth). We generally are restricted to collecting data from a sample, or subset of the population, and use statistical procedures to estimate what we're interested in knowing about the population. For instance, we might be interested in whether two different strains of bacteria have different colony growth rates. We don't test all bacteria but, instead, have samples of colonies that we hope represent the larger population of all bacterial strains. For these types of questions we are not just summarizing numerical information but are, instead, asking what these samples of data mean in the larger context. This extension of a hypothesis from samples to statements about populations is referred to as "inference."

In this way we generally collect data on a small subgroup and extrapolate what it means to the larger group. Our small group is referred to as a *sample* while the larger group is called a *population*. Summary values of a sample are called "statistics" while summary values of the entire population are called "parameters." We, therefore, usually collect "samples" from "populations" to get "statistics" that we use to estimate population "parameters." The process of understanding populations from samples using a variety of statistical procedures is referred to as "inferential statistics." Much of what we will do in the rest of this book is designed to help you make this leap from summary statistics to inferential statistics.

6.2 HOW TO ASK AND ANSWER SCIENTIFIC QUESTIONS

Here are the basic steps that scientists take when they want to know something about the universe that can't just be looked up. In general, these steps are really hard to accomplish. Scientists usually need extra funding to do this work and so scientists work to get grants which fund the research. Biological research, in particular, is expensive and time consuming. It is, therefore, quite important that the data be collected carefully and analyzed correctly. Your success in learning to design experiments and analyze the results provides you a valuable and marketable skill. So, here are the steps scientists tend to follow, often referred to as the "scientific method":

1. Clearly state your question in a way that could be tested. This question is generally stated as a "hypothesis."
2. Decide what data are needed to answer the question. These should be just the data you need to answer the question—no more and no less.
3. Hand draw a graph of what a beautiful answer might look like if your idea is correct. This exercise will always help you clarify your experimental design.
4. Determine the appropriate analysis that would be required to test your hypothesis if you got the cool data you graphed by hand. You should MAKE UP DATA (called "dummy data") and try the analysis. What???? Yes! This way you really understand what you're trying to find out and whether the data you hope to collect will answer your question.
5. Actually do your statistical test on the dummy data. Evaluate your result. Consider performing a power analysis (see section 6.4 on page 92) using these data or data from previous experiments or the scientific literature.
6. Design a good way to get those data (experiment or observation). You will likely draw heavily on what others have done before.
7. Do the research and collect the data as carefully as possible.
8. Look at the data you have collected. This is best done with a graph. Do the results look like your hand-drawn graph? Test how the data are distributed (normal or not normal).
9. Test your hypothesis. This is where inferential statistics are used.
10. Share your result with the greater world through a carefully crafted report (e.g., see the journals *Science* or *Nature*) or presentation. Use the style required by your reader (e.g., lab instructor or professor). It doesn't matter how strange that might seem. The only thing that matters is

that you follow the formatting requirements. Different journals have different styles (as do different faculty!).

Be sure to respect the investment you've made into getting each data point. Time, lab equipment, consumables (e.g., reagents), and the time of faculty, lab instructors, and assistants are all costly. This leads us to an important rule:

"The Golden Rule of Data Analysis"

You should be as careful conducting your data analysis and developing your visualizations as you are designing the experiment and collecting the data. Your results are only as valid as the weakest part of your research.

6.3 THE DIFFERENCE BETWEEN "HYPOTHESIS" AND "THEORY"

People often assume that a hypothesis is just a fancy word for a guess in science. This is not correct! A guess is generally some estimate that is made with insufficient evidence or an understanding of the system. For instance, I am an avid tennis player and before a match players spin a racquet to decide who gets to choose whether or not to serve (e.g., "up" or "down"). The opponent has no information beyond "up" or "down" and we assume that there is an even probability of getting either outcome. The person who gets to choose up or down is making a guess. So, what's a "hypothesis"?

A scientific hypothesis

A hypothesis is a well-educated prediction of an outcome that would occur from a scientific experiment.

In general, when we've gotten to the point of developing an experiment we are testing a very well thought out hypothesis that has been developed based on previous research experience and/or a deep exploration of the scientific literature. We're never just guessing as to what might happen–experiments are too costly! Ultimately, we test hypotheses statistically, which usually provides us with some level of confidence, or lack of confidence, in one of our hypotheses. And for our purposes in this book we are generally testing the "null hypothesis" (H_0). The H_0 is one of many potential outcomes and usually agrees with random chance, no pattern, or no relationship. When we do experiments we often are interested in positive results (e.g., something is happening different from chance) so we test whether our data support the H_0 or not. We then report our result and include a "p-value." We will develop the technical definition of the p-value very carefully below (section 6.6).

A related term we often hear is "theory." I once had a mathematician friend of mine state confidently that evolution isn't true–it is "just a theory." He even noted that scientists refer to it as the "theory of evolution." For evolution to be true must we prove that it is true? Otherwise, isn't it simply a guess? Are scientific theories just guesses? Where's the proof?

The bad news is that we do not prove anything in science. Mathematicians prove stuff and they can do this because, well, they cheat by building proofs that rely on axioms which are *assumed* to be true. In biology we work to understand nature and continue to challenge our ideas with more evidence. Evidence, however, is always a bit shaky and dependent on many *variables* and sources of *error*. As we work to understand how nature works we gather more and more evidence that supports or refutes ideas. If the ideas are not supported by data we may modify our understanding of nature or possibly abandon our ideas. So, what's a "theory"?

A scientific theory

A theory is a comprehensive explanation of natural phenomena supported by extensive evidence gathered through observations and/or experiments.

The evidence we gather may increase our confidence that our ideas about nature are correct. But, unfortunately, we only get more and more confident. We do not ever eliminate the possibility that some new evidence might cause us to have to give up on our beloved ideas. We, therefore, cannot be 100% sure an idea is true which, therefore, prevents us from "proving" anything in science. A slightly controversial corollary to this is that, if we can't prove something true in science, we can't disprove it, either. If we can't be 100% sure that something is true, then we also can't be absolutely certain that something is false.

What then can we say about our results? We can't say: "our results *prove*" that something is happening. Instead, we say that "our results are *consistent*" with something happening.

So, the words "prove" and "disprove" have limited roles in science, restricted particularly to the area of applied mathematics. Biomathematicians, for instance, do provide proofs that lead us to deeper understandings of how biological systems work, or can work. Therefore, our inability to prove or disprove hypotheses in science is the result of the inescapable fact that ***all results in science are provisional***.

6.4 A FEW EXPERIMENTAL DESIGN PRINCIPLES

As we begin to think about analyzing data it is good to know how we got the data and, if required, how to get data. Here are a few design principles to keep in mind. Most, but not all, good experiments will include the items in this list. We should accept, however, that occasionally we can gain great insight about nature without some of these components. Some biological systems, for instance, are just too large and difficult to replicate (e.g., atmospheric response to elevated CO_2 or lake ecosystems).

1. **Replication**. This solves the problem that we might have if we only measured one individual who happens to be really unusual. We need, instead, to sample a bunch of individuals under each treatment level so that we can understand the collective behavior of our system. Each individual should be *independent* from other individuals. It is possible, however, to measure a single individual multiple times but we need to be careful. For example, we might measure the mass of an individual over time to determine its growth rate, but we only get one growth rate measurement from this individual and, therefore, need to make the same type of measurement on other individuals (replicates). A problem might arise if all our study critters are genetic clones within a species. Are these independent replicates? In general, our inference will only extend to this clone, not the species as a whole.

2. **Randomization**. We often need to take individual samples in a study and place them in different treatment groups randomly. If we're sampling out in nature we need to pick our sites randomly. It's not random, for instance, if we choose research plots that are close to a road. If we do this, then we only can say something about the forest next to the road, not about the forest in general (see section 6.5 on page 93).

3. **Factors**. Factors are different treatments that we're investigating that might affect our study system. For plants, we might be interested in the effect of nitrogen on growth rates. We can have multiple factors that we investigate (e.g., water and temperature).

4. **Levels**. If we have a factor, then we need to have at least two different *levels* of that factor to understand how it influences our study system. In the above example, we might investigate the effect of nitrogen on plant growth rate and so we have a level with no nitrogen added (*control*) and a level with nitrogen added.

5. **Control**. Controls are tricky sometimes. In the above example, the "no nitrogen" treatment serves as the control. We like to have controls that are factor levels against which we compare our treatments. If, however, we are interested in testing whether SAT scores for males and females are not different, then neither gender would be considered to be a control group.

INFERENCE

Once we have designed a study and collected and analyzed our data, we usually want to say something about the natural world as a whole. This statement of the greater meaning of our research is referred to as "inference." Our ability to make inferential statements is limited by the data we have collected. If, for example, we are interested in endocrinology and have discovered the function of a hormone on metabolic activity in mice, what can we say? We might be tempted to state that our result is important in mammals. We might even go as far as saying that this could be important to humans. Unfortunately, if we've studied this in mice we can't say this result applies even to related mammal species—and certainly not to humans.

Studies using "model" systems, such as fruit flies, mice, or yeast often are done using just one or a few genetic lines where all individuals within a line are genetically identical. In this case, our inference (statement of the greater meaning of our result) can not go beyond this line of mice. If we use three strains of white lab mice and the result was found clearly in all three, can we say something about this result in white lab mice in general? Possibly, if these three strains were chosen *randomly* out of all strains of white lab mice. Most likely, however, these mice strains were not chosen randomly but were, instead, strains with which the investigator was familiar or which are simply particularly easy to raise and test under laboratory conditions. Therefore, inferential statements might not go beyond the three specific strains of mice.

Think of it this way: if you discover something about three of your friends, are you ready to make an inference about humans?

POWER ANALYSIS

If we're designing an experiment, we always have constraints on the size of our experiment. We just discussed the importance of replicates. Why not just have as many as we can afford, have room for, and/or have personnel to do the work? Basically, because we will be making one or more of a variety of mistakes if we do this. We want to be efficient and design the right experiment for the hypothesis we are testing. If we use too many resources, we will not be able to ask other questions. If we use too few resources, then we might not be able to answer our question.

One solution to this problem is for us to conduct a "power analysis" or a "power test." The idea of this is to estimate the number of replicates we need to determine that a factor (e.g., nitrogen) affects our response variable (e.g., plant growth rate). The problem with this determination is that we kind of need to know the answer before we start!

As an example, let's assume that a t-test is the appropriate way to determine if the masses of individual rodents are different between two geographically isolated morphotypes within a species. From previous research we know that the mean mass of one morphotype is 250g and that the variability for this species is $s = 25$g. Given this information, can we estimate how many animals we would have to collect and weigh to detect a significant difference of at least 30g? The "power" of a statistical test is defined as $1 - \beta$, where β is the probability of making a Type II error (see section 6.7). Determining power is difficult and depends on sample size, our critical p-value α, the strength of the effect we're studying, and various errors we might make. Often a value of 0.1 is a conservative estimate ($\beta = 0.1$, power $= 1 - \beta = 0.9$). Given this, our power analysis is done as follows:

```
> power.t.test(delta = 30, sd = 25, power = 0.9)

    Two-sample t test power calculation

            n = 15.61967
        delta = 30
           sd = 25
    sig.level = 0.05
        power = 0.9
```

```
alternative = two.sided
```

```
NOTE: n is number in *each* group
```

This suggests that we need at least 16 replicate individuals (always round up) in each sample group to have our t-test provide us with a significantly different mass for these populations, given that we need the means to be different by 30g, that the standard deviation of these samples is 25, and that our samples are normally distributed (see chapter 4). This function has more flexibility and other power tests are available.

6.5 HOW TO SET UP A SIMPLE RANDOM SAMPLE FOR AN EXPERIMENT

Let's imagine that we are asked to set up an experiment to test the effect of gibberellin on the height of *Brassica rapa* (a fast-growing plant). We would like to employ a proper scientific approach for this experiment so that we can quantify and interpret our results correctly. Let's apply what we learned in the previous section on the principles of experimental design.

1. **Replication.** To test the effect of gibberellin on plant height, we need to be sure that we have several plants. Our laboratory instructor might provide us with 10 pots for our replicates.

2. **Randomization.** In each pot we overplant them with seeds and, during the first week, we carefully remove all but one plant per pot. At the start of our second week we are prepared to apply our treatment (spray gibberellin onto the leaves of the treatment plants). Which plants receive the treatment and which do not? We should do this randomly using the approach described in section 6.4 rather than haphazardly (just choosing them ourselves).

   ```
   > sample(1:10,5) # from the array 1-10, randomly choose 5

   [1] 3 6 4 7 2
   ```

   If you run this line you should get a different set of plants than I did. These are the pots to spray with gibberellin. The other pots do not receive the treatment. See also the sampling section 3.5 on page 41.

3. **Factor.** We have just one factor in this experiment: gibberellin.

4. **Levels.** For our gibberellin factor we have just two levels: with and without gibberellin applied.

5. **Control**. The control plants are those that do not receive gibberellin. In this example, gibberellin is sprayed on the plant leaves at the beginning of week two. Our control should be a spray that doesn't contain gibberellin, such as distilled water.

6.6 INTERPRETING RESULTS: WHAT IS THE "P-VALUE"?

A p-value is an important number that comes from conducting many of the statistical tests discussed in this book. It's usually a little tricky to understand at first so let's start slowly. When you conduct one of many statistical tests you often get a p-value that tells you something about your hypothesis. A simple test might ask whether two samples are statistically different from each other. What we are generally asking is whether the two samples came from the same population. A simple example might be to test whether gibberellin increases plant height. If the samples are both normally distributed (an assumption of the t-test), then you can perform a t-test (see section 7.3). The test returns a p-value (see Figure 6.1). In a t-test the t-value generally gets larger as the means of the two samples get more and more different (but this also depends on how messy the data are). Larger t-values generally yield smaller p-values and lower our confidence that the two samples came from the same population.

As a first approximation (because it's not technically correct), the p-value might be thought of as the probability that the null hypothesis is true. With this definition we might think that a large p-value, like 0.95, is good support for the null hypothesis, while a p-value of 0.001 provides very little support for the null hypothesis.

A correct definition is shown in the box below.

The p-value

The p-value is the probability of getting a test statistic (like t returned by a t-test or F returned by an ANOVA test) that is as extreme or more extreme than the one you observed, had the null hypothesis been true.

I assume this definition seems pretty confusing. It's not that bad, however. Statistical tests return "test statistics" (e.g., a t-value or an F-value). These test statistics have ranges where, at one end, they suggest there's no difference or relationship. At the other end, the test statistic might suggest

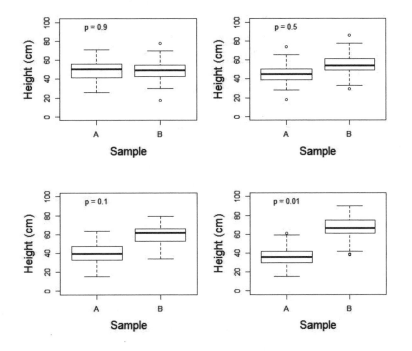

Figure 6.1: Four potential outcomes from a study. In the upper-left panel it appears that the two samples came from the same population (p = 0.9). As the samples appear more and more different our p-value from a t-test gets smaller. At some point our samples appear so different that we conclude that they did not come from the same population. We often draw this line at the critical p-value called α. In general, we set $\alpha = 0.05$.

a big difference between samples or a strong relationship between samples. Generally, when these test statistics are near zero we *fail to reject the null hypothesis*. Large test statistics reduce our confidence in the null hypothesis. So, for the t-value, for instance, it continues to get larger for samples that are more and more different and, at some point, we reject the null hypothesis that the samples came from the same population (see Figure 6.1).

Our definition of a p-value above explains why we consider a p-value that is "less than or equal" to the critical p-value (α) to be "statistically significant." We generally set $\alpha = 0.05$ prior to conducting an experiment that tests a null hypothesis (H_0). We, therefore, reject the null hypothesis when $p \leq \alpha = 0.05$.

Notice, too, that we are dealing with statistical significance while biological significance is a "whole nother animal."

Where did $\alpha = 0.05$ come from? That's a pub story involving a famous geneticist named Sir Ronald Fisher. It is an arbitrary value. But, it's a widely accepted value. We will use the convention that we fail to reject the H_0 when $p > \alpha$. Not everyone will agree with this rule. But it is a simple rule for getting started. Therefore, if $\alpha = 0.05$ and $p = 0.06$ it is wrong to say something like "the relationship is *almost* significant" or that "there is a positive *trend*." The finding is simply something like this: "The samples are not significantly different (t = ___, df = ___, p = 0.06)."

So, can you see the correspondence between our definition of the p-value and the comparison of a p-value to α? As we work through a variety of tests in chapters 7–11 you will gain confidence in interpreting the results from experiments.

6.7 TYPE I AND TYPE II ERRORS

In science we cannot be absolutely certain of anything unless we're doing mathematics. In biology, if we're dealing with data, there always is a chance that we are wrong. This is true even for physicists, at least if they're working with data using a natural system. We accept that there is a probability of making a mistake and have to balance this chance with our desire to correctly interpret our results. Why not just make α really small so that if we reject our H_0 we are really sure the H_0 is wrong? This is a great question because it turns out that if our H_0 is really true then we will mistakenly reject this with a probability of α. So, why not just make $\alpha = 0.001$ and we'll rarely make this mistake—erroneously reject a true H_0?

Let's consider that we really don't want to make this mistake of incorrectly rejecting a true H_0. For context, let's imagine we're investigating whether human males and females are different in height. We set $\alpha = 0.001$ so that we'll only reject the H_0 (no difference in heights) if the heights of males and females are *really* different. To see a significant difference, then, the samples might have to have a difference of, perhaps, three feet! If the samples were that different then we'd be pretty certain males and females have different heights and we would be less likely of incorrectly rejecting the H_0. More technically, we've made the chance of making a Type I error very low because it's really unlikely we'd reject this H_0. But, unfortunately, under this scenario, if males and females were different by *only* 2.5 feet then we'd conclude that height of males and females are not statistically different! We would likely agree we're making a mistake!

This becomes more problematic when the results matter, such as in the case of a new drug that could save or extend lives. If the life expectancy of patients increased because of our drug, what length of time might be considered "significant"? We are likely to think one week might be great. If we set α to a small value we would not likely detect such a difference. We would be really rigorous in our result by wanting to be *certain* that our drug extends life by a long time. If our drug does extend life but, perhaps, not by much, or the result is pretty variable, we *fail to reject* our false H_0 (Type II error, see Figure 6.2). We might end up not recommending that a drug that works be administered. Are you comfortable with that?

To avoid this mistake we might set $\alpha = 0.2$. This is great because the test is really sensitive. We would be more likely to reject our H_0. So what? This error might be acceptable because maybe the drug is inexpensive and has a placebo effect so, even if it doesn't work and we make a Type I error (incorrectly reject a true H_0), no one is hurt (see Figure 6.2). Sometimes we must consider which of these errors is more acceptable!

6.8 PROBLEMS

(For questions 1–3). You perform a statistical test and get a p-value of 0.06.

1. Under what conditions would we consider this to be statistically significant? How about not to be statistically significant?
2. What might we mean by *biological* significance?
3. Assuming the researcher set $\alpha = 0.05$ prior to conducting the experiment, describe the *type* of error made by concluding the p = 0.06 is close enough to being signficant so as to reject the H_0.

4. You walk into your bedroom and flip on the light switch. The light does not go on. Assuming this is a problem, describe the steps that you would take to remedy this using the scientific method.

(For questions 5–6). You plan to investigate whether three levels of watering rates affect the height of a plant species grown from seed to two months of age. The levels are 50 ml, 100 ml, and 200 ml per day.

5. Provide a hand-drawn graph of what the results might look like.
6. Describe what the control is (or should be) in this experiment.

What we do

	Fail to reject H_o	Reject H_o
H_o Is really true	Correct	Type I Error P(I) = α ("false positive")
H_o Is really false	Type II Error P(II) = β ("false negative")	Correct

Figure 6.2: This diagram represents Type I and Type II errors. If we could assume that our null hypothesis (H_0) is true, then, based on our data analysis, we are acting correctly if we fail to reject the H_0. If we reject a true H_0, however, then we've made an error (Type I). If, however, we fail to reject a false H_0, then we've also made a mistake (Type II). Finally, if we reject a false H_0, then we've acted correctly. Note the probability of making a Type I error is α while the probability of making a Type II error is β.

7. Three student researchers are interested in how a particular plant species responds to elevated atmospheric CO_2. They clonally propagate cotton-wood saplings (*Populus deltoides*) so as to reduce the variability among individuals which, therefore, maximizes their ability to attribute any response they see to the CO_2 treatment effect. What inference can they make about plant response to changes in CO_2?

(For questions 8–10). Below is a quote from a *New York Times* (Sept. 12, 2012) article titled "How testosterone may alter the brain after exercise."

A new study published last month in *Proceedings of the National Academy of Sciences* [found] that male sex hormones surge in the brain after exercise and could be helping to remodel the mind. The research was conducted on young, healthy and exclusively male rats—but scientists believe it applies to female rats, too, as well as other mammals, including humans.

8. Describe the population that these results apply to.
9. Describe why the results can or can't apply to female rats.
10. Describe why the results can or can't apply to humans.

CHAPTER 7

HYPOTHESIS TESTS: ONE- AND TWO-SAMPLE COMPARISONS

This chapter tackles hypothesis testing when we have one sample and a test value or two samples that we want to compare against each other. Choosing which test to use can be tricky. I've tried to group tests together based on the structure of your data. However, which test you use may be influenced by experience and by suggestions of your laboratory instructor or research mentor. If you are not sure what to do, keep in mind you've learned a lot and should be able to ask a qualified person for directions and implement the solution with one of the techniques introduced in this and the next two chapters. The tests presented here form a starting point for you. No matter your questions, however, the answers can be gotten with R so you're definitely on your way to solving any problem in biology.

7.1 TESTS WITH ONE VALUE AND ONE SAMPLE

These are relatively simple tests since there's only a single sample against which we're testing a single value. The sample may or may not be normally distributed. We also may or may not assume directionality in our test (e.g., the sample is greater than some value).

Sample is normally distributed

Perhaps you are interested in knowing whether your combined SAT score of 1330 is not statistically different from a randomly drawn sample of students from your college. You can enter the sample of student SAT scores like this:

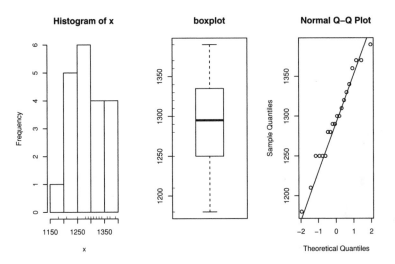

Figure 7.1: Assessing the SAT scores for a sample of college students using the `simple.eda()` function. The plots seem to suggest that the data are consistent with a normal distribution.

```
> S = c(1370, 1290, 1250, 1320, 1280, 1250, 1370, 1340, 1330,
+       1280, 1210, 1390, 1180, 1290, 1300, 1300, 1250, 1250,
+       1310, 1360)
> you = 1330
> simple.eda(S)
```

First, you should visualize your data, specifically the SAT sample from the group of college students. This can be done using the `simple.eda()` function discussed in section 4.4 (see Figure 7.1). Given this, it seems plausible that this sample is normally distributed. Next, you should test whether the sample is normally distributed. Note that our dataset is small and we should, therefore, be a little skeptical. We can ask R whether the data appear to be normal using the Shapiro-Wilk test. Below I ask R to simply report the p-value to save room.

```
> shapiro.test(S)$p.value
```

```
[1] 0.8485299
```

We don't have enough evidence to be concerned about our assumption of normality (see section 4.4) so we can proceed with a t-test. Note that we are just asking whether there is a difference between your 1330 SAT score and the scores from a sample. We only have one sample and, therefore, call this is a "one-sample t-test."

```
> t.test(S, mu = you) # mu value tested against sample

        One Sample t-test

data:  S
t = -2.7597, df = 19, p-value = 0.01247
alternative hypothesis: true mean is not equal to 1330
95 percent confidence interval:
 1270.213 1321.787
sample estimates:
mean of x
     1296
```

Above is the output from the t-test. Wow, there's a lot of stuff! The important part is the second line of the output which suggests that the sample of SAT scores from your friends is statistically different from your SAT score (t = -2.76, df = 19, p = 0.0125). We can graph the result as a boxplot and add a reference line representing the test score of 1330. This is a good way to show the relationship between a sample and a test value (see Figure 7.2).

```
> boxplot(S, ylab = "Combined SAT Scores", cex.lab = 1.5)
> abline(h=you, lty = 2, lwd = 2) # horiz. ref. at score=1250
```

Your score is higher than this sample and it is statistically higher, as well (t = -2.76, df = 19, p = 0.0125). I'm going to revise my result statement so, instead of saying the sample is statistically different, I can actually state the result. Since the sample of friends is different and their sample mean is lower I know that the score of 1330 is statistically higher than the sample of students. The question had no directionality, which means it's what's called a "two-tailed test." That's because we were looking to see if your SAT score fell into or beyond either the upper or lower tails of the sample distribution. In answering our question, however, I've added directionality because there is no other conclusion I could make (your SAT score is not below the sample

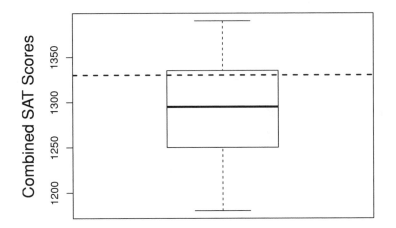

Figure 7.2: A boxplot of the combined SAT scores for 20 randomly chosen individuals. The dashed line, added by sending the argument `lty = 2` to the `abline()` function, is the reference line for your SAT score of 1330.

mean) and I want to leave my reader with the most complete result. I also have to give my reader the graph (see Figure 7.2).

What if instead I had already guessed that your SAT score was high, compared to the average students at your college? This is a different hypothesis; this is a *directional* hypothesis. In statistics we call this a "one-tailed test" because we're asking if the test value is in or more extreme than one of the tails of the sample distribution. Such a question would sound something like this: "Is your SAT score statistically higher than a random sample of students from your school?" To test this we focus on the *sample* and not the individual value. So, even though we're testing if your SAT score is greater than our sample, our statistical test evaluates whether the *sample* is, in this case, *lower* than our test value (your SAT score). The test is then conducted with this code:

```
> T = t.test(S, mu = you, alt = "l") # alternative hypothesis
>                          # is that the sample is greater than mu
> T$statistic # report the t-value

        t
-2.759676

> T$p.value # report the p-value

[1] 0.006235234
```

Note that the t-value has not changed but that the p-value is smaller than our previous 0.0125 (it's half of it). This result suggests that the sample is statistically less than the test value (1330). Your SAT score is significantly higher than the sample (t = −2.76, df = 19, p = 0.0062). The lower p-value is due to the increased power that we get from narrowing our hypothesis test. Both tests (one- and two-sided, one-sample tests) yield a rejection of the null hypothesis. To make matters confusing, our conclusion is the same for both. You can probably guess that there's a window where this might not be true. Which test do you use? You should **always** use the test that matches your hypothesis. Therefore, you always need to be careful in how you phrase your hypothesis. Here are the two hypotheses we have tested. [Note that under all circumstances you test only one hypothesis: the correct hypothesis for the question you are trying to answer. I sometimes conduct a few hypotheses for explanatory purposes.]

1. Your SAT score is not statistically different from the sample of scores (no directionality so it's a two-tailed test).
2. Your SAT score is less than or greater than the sample of scores (there is directionality so these are one-tailed tests).

In R we provide the alternative hypothesis to our call to the `t.test()` function by adding the argument `alt = "g"`, where `alt` stands for alternative and `g` refers to the sample being greater than our test value. If we had hypothesized that the sample was less than your SAT score, then we would have used `alt = "l"` (that's the letter "el"). Based on this, our alternative hypothesis can be written as $H_0 : \bar{x} > \mu$ with $\bar{x}$ representing the mean of the sample and μ representing our test value. All of our hypotheses must be "mutually exclusive and all inclusive," so our null is written as $H_0 : \bar{x} \leq \mu$. [Note that I have shown two statistical tests performed on one dataset. In practice this is *never* appropriate. You simply state the hypothesis you are interested in testing and then test just that one hypothesis!]

Sample is not normally distributed

This is the same question as above but your friends, refreshingly, are not "normal."

```
> friends = c(1340, 1540, 1310, 1270, 1340, 1300, 1330, 1540,
+               1540, 1320)
> you = 1300
> shapiro.test(friends)

        Shapiro-Wilk normality test

data:  friends
W = 0.7477, p-value = 0.003344
```

From this we see that the sample is not normally distributed ($p \leq 0.05$). Therefore, we should use the non-parametric Wilcoxon test. Here we're asking the simple question as to whether the sample of friends have SAT scores that differ from your score.

```
> wilcox.test(friends, mu = you) # mu tested against the sample

        Wilcoxon signed rank test with continuity correction

data:  friends
V = 41.5, p-value = 0.02757
alternative hypothesis: true location is not equal to 1300
```

We conclude that your SAT score is statistically lower than the sample (V = 41.5, df = 9, p = 0.028).

7.2 TESTS WITH PAIRED SAMPLES (NOT INDEPENDENT)

The difference between pairs is normally distributed

As we will see, with paired data we are interested in the difference between each paired value, thus the title of the section. Below are data for the length of time it takes six chimpanzees to solve a problem. The "before" measurements are minutes for a naive chimp to solve the problem while the "after" times are minutes taken after training.

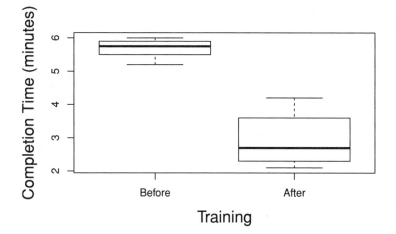

Figure 7.3: An incorrect representation of the length of time taken by chimpanzees to solve a problem. This graph suggests that there are two independent samples of data for the chimps. Unfortunately, the chimps in both samples are the same.

```
> before = c(6.0, 5.8, 5.7, 5.5, 5.9, 5.2)
> after = c(2.1, 3.6, 2.8, 2.6, 4.2, 2.3)
```

There are various ways to visualize these data. One way you might consider is shown in Figure 7.3. This was done using the following code:

```
> boxplot(before,after, names = c("Before","After"),
+       xlab = "Training", ylab = "Completion Time (minutes)",
+       cex.lab = 1.5)
```

Why might this not be the best way to visualize these data? It seems to do a nice job showing the distribution of response times before and after and we see that completion time for the task went down after training. The problem is that this graph suggests that we have two independent samples. We don't! It's the same reason we don't test the variables "before" and "after" for normality. We only have a single measure of improvement from each chimp, or *experimental unit*. Therefore, the proper way to represent these data is with either a single boxplot (Figure 7.4) that shows the difference from zero or a bump chart that shows the reduction in completion time of each chimp (see Figure 5.8 on page 79 for the bump chart used on a different dataset).

It's important to keep in mind, also, that our test for normality should be conducted on the single sample distribution represented by the difference between before and after (the variable `diff`), as follows:

```
> diff = after - before # create a single, new variable of diffs
> shapiro.test(diff) # test this single sample for normality

        Shapiro-Wilk normality test

data:  diff
W = 0.9286, p-value = 0.5693
```

Here's how we can now proceed with the graphing of these data (see Figure 7.4).

```
> boxplot(diff, ylim = c(-4,1), cex.lab = 1.4,
+         ylab = "Change in Learning Time (min.)")
> abline(h = 0, lty = 3, lwd = 3)
```

The question was whether *improvement* occurred because of the training. We're specifically asking a directional hypothesis–whether chimps improved in the amount of time needed to complete a task (if $time_{after} < time_{before}$). We can rewrite this hypothesis as $time_{after} - time_{before} < 0$. [Note that $time_{after} - time_{before} = $ `diff`.] Alternatively, there was no improvement or they even got worse with training. This, it turns out, is our H_0 since it includes no effect. This can be written as $time_{after} \geq time_{before}$ or, by subtraction $time_{after} - time_{before} \geq 0$ (or `diff` ≥ 0). Therefore, since our samples are not independent but rather repeated measures on the same individuals, the correct test is a paired t-test that asks whether the amount of time taken after training ("after") was *less* than the time required to complete the task before training. Note that, in this test, we supply our *alternative hypothesis* to R.

```
> t.test(after, before, paired = TRUE, alt = "l")

        Paired t-test

data:  after and before
t = -9.0095, df = 5, p-value = 0.0001406
```

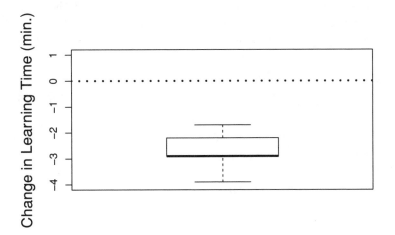

Figure 7.4: Data from Figure 7.3 graphed correctly. This graph shows the reduction in task completion time as a result of learning. The reference line at zero represents no change and is well outside our distribution. This line shows the reader what we would expect if there was no improvement due to learning.

```
alternative hypothesis: true difference in means is less than 0
95 percent confidence interval:
     -Inf -2.134942
sample estimates:
mean of the differences
                -2.75
```

The test returned by R is of the H_0, which was $time_{after} \geq time_{before}$. The small p-value ($p < 0.001$) suggests we reject the H_0. Therefore, the training of chimpanzees reduced the amount of time required to complete the task (t = −9.0, df = 5, p < 0.001).

Difference between paired data is not normally distributed

The LDL cholesterol levels of individuals were measured for 10 human subjects. These individuals were then "treated" by restricting them to meals served only at an unnamed fast-food restaurant for 30 days. Is there a change in LDL for these individuals? Note that we're just asking if there's a difference, not whether there was an increase in LDL levels. Therefore, this is not a directional test. Here are the data:

```
> before = c(75, 53, 78, 89, 96, 86, 69, 87, 73, 87)
> after = c(124, 103, 126, 193, 181, 122, 120, 197, 127, 146)
```

Since we only are interested in the changes, we test whether those differences (`after-before`) are normally distributed:

```
> shapiro.test(after-before)$p.value
```

```
[1] 0.02916288
```

The p-value for the Shapiro-Wilk test is less than 0.05 so we reject the null hypothesis that the data are normally distributed. Therefore, we proceed with the non-parametric Wilcoxon test.

```
> wilcox.test(before,after,paired = TRUE)

        Wilcoxon signed rank test

data:  before and after
V = 0, p-value = 0.001953
alternative hypothesis: true location shift is not equal to 0
```

We note that there is a significant difference in LDL levels before and after the treatment. In particular, the LDL of our sample of individuals increased significantly after one month of eating at a fast-food restaurant (V = 0, df = 18, p = 0.002). The graphing of these data would follow the same format that we saw for the paired t-test. [Note that no individuals were actually hurt to get these data.]

7.3 Tests with Two Independent Samples

In the previous section we looked at examples where there was only one sample. In this section we'll look at tests where there are two, independent sample groups.

SAMPLES ARE NORMALLY DISTRIBUTED

Imagine we have two samples and we want to know if they're different or, more technically, if the two samples came from the same population. Our null hypothesis is that they did come from the same population and are, therefore, not statistically different ($H_0 : \bar{x}_1 = \bar{x}_2$). Here are the heights of 10 plants grown with fertilizer (`fert`) and 10 plants without (`cont`).

```
> cont = c(64.7, 86.6, 67.1, 62.5, 75.1, 83.8,
+ 71.7, 83.4, 90.3, 82.7)
> fert = c(110.3, 130.4, 114.0, 135.7, 129.9,
+ 98.2, 109.4, 131.4, 127.9, 125.7)
```

We should always begin by visualizing our data. A side-by-side boxplot is an appropriate approach (Figure 7.5).

```
> boxplot(cont,fert, names = c("Control","Fertilizer"),
+          xlab = "Treatment", ylab = "Plant Height (cm)",
+          cex.lab = 1.5)
```

This graph seems to suggest that fertilizer increases plant height. We also see that the two samples might be normally distributed. Let's test them for normality. Since these two samples are independent we need to test them individually. Here are just the p-values from the normality test:

```
> shapiro.test(cont)$p.value # normal?
```
p, normal tests Ho or H$_A$

```
[1] 0.3725881
```

```
> shapiro.test(fert)$p.value # normal?
```

```
[1] 0.1866711
```

The samples appear to be normally distributed so we may proceed with a standard t-test. The t-test also assumes that the variances are equal. Let's test that, too, using the `var.test()` variance test.

```
> var.test(cont,fert)
```

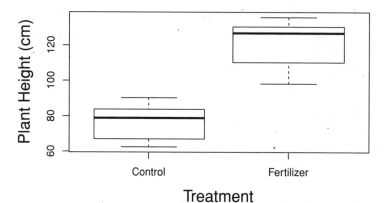

Figure 7.5: A comparison of the heights of plants with fertilizer and without (control). The fertilized plants are significantly taller than the controls (t = −8.884, df = 18, p < 0.001). Do the data in the graph support this statistical finding?

```
        F test to compare two variances

data:  cont and fert
F = 0.638, num df = 9, denom df = 9, p-value = 0.5137
alternative hypothesis: true ratio of variances is not equal to 1
95 percent confidence interval:
 0.1584741 2.5686485
sample estimates:
ratio of variances
         0.6380159
```

The resulting p-value (p = 0.51) suggests that we should not reject the H_0 (variances are not different). Therefore, we should conduct a t-test instead of the Welch test (done if variances are unequal). We are now able to complete our t-test, adding the argument **var.equal = TRUE** to our function call:

```
> t.test(cont,fert,alt = "l", var.equal = TRUE)
```

```
        Two Sample t-test

data:  cont and fert
t = -8.884, df = 18, p-value = 2.67e-08
alternative hypothesis: true difference in means is less than 0
95 percent confidence interval:
       -Inf -35.81405
sample estimates:
mean of x mean of y
    76.79    121.29
```

From this we find that fertilizer significantly increased the height of plants (t = −8.884; df = 18, p < 0.001, see Figure 7.5).

SAMPLES ARE NOT NORMALLY DISTRIBUTED

In another experiment, we might test whether the heights attained by two plant species at a certain time are statistically different (note the lack of directionality in our hypothesis). Here are the data for heights in cm:

```
> A = c(87.1, 86.4, 80.3, 79.4, 85.7, 86.6, 70.5,
+       80.8, 86.3, 70.7)
> B = c(103.9, 103.6, 109.7, 108.6, 102.3, 103.4,
+       119.5, 109.2, 103.7, 118.3)
```

We should first visualize these data. A boxplot would be appropriate:

```
> boxplot(A,B, names = c("Species A","Species B"), las = 1,
+         ylab = "Plant Height (cm)", cex.lab = 1.5)
```

The data do not appear to be normally distributed. We, however, need to test these samples for normality.

```
> shapiro.test(A)$p.value # normal?
```

```
[1] 0.01783138
```

```
> shapiro.test(B)$p.value # normal?
```

```
[1] 0.02395352
```

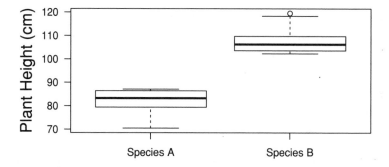

Figure 7.6: A comparison of heights between two different plants grown for two months. Note that we performed a two-tailed, non-parametric Wilcoxon test on these data. Based on these boxplots do the samples appear normally distributed?

After determining that the samples do not appear to have come from normal distributions we should proceed with a non-parametric Wilcoxon test.

```
> wilcox.test(A,B)


        Wilcoxon rank sum test

data:  A and B
W = 0, p-value = 1.083e-05
alternative hypothesis: true location shift is not equal to 0
```

If we instead had wanted to conduct a test with a directional hypothesis then we should have proceeded as we did in section 7.1 on page 101. From this analysis we conclude that the heights of individuals of species B were statistically greater than those of species A (W = 0, df = 18, p < 0.001, see Figure 7.6).

7.4 PROBLEMS

(For questions 1–4). Below are the number of adult smaller tea tortrix moth (*Adoxophyes honmai*) counted in different light traps (Nelson et al., 2013).

1916, 1563, 1436, 6035, 3833, 5031, 13326, 3130, 6020, 1889

1. Create an appropriate visualization of the distribution of these data. Do the data appear normally distributed?
2. Test whether the data are normally distributed.
3. If we collect new data for 2014 and find 20,000 adults, would this be statistically unusual?
4. Create a visualization of these data, including the value of 20,000 adults that might have been observed in 2014.

(For questions 5–6). The order of leaves along the stem of plants is referred to as phyllotaxis. A researcher is interested in whether the 5th order leaves (lf5) differ in leaf area from the 1st order leaves (lf1). The leaf areas, in cm^2, for six plants are shown below.

$$lf1 = 27, 26, 20, 24, 28, 29$$
$$lf5 = 30, 34, 28, 35, 42, 40$$

5. Are these leaf areas different?
6. Create the most appropriate visualization of your results.

(For questions 7–8). Two unnamed universities (U1 and U2) compete in a championship basketball game. The height of the 12 players on each team are listed below, in inches. Sports pundits say height matters in the game.

$$U1 = 81.0, 80.1, 86.1, 78.9, 86.8, 84.6, 79.3, 84.0, 95.4,$$
$$70.3, 86.8, 78.1$$
$$U2 = 94.4, 76.7, 70.0, 88.8, 73.7, 86.3, 85.7, 74.0, 79.5,$$
$$75.9, 68.1, 75.9$$

7. Is there enough evidence to suggest one university has an advantage over another? If so, which has the height advantage?
8. Create an appropriate visualization of your result.

(For questions 9–10). Below are combined SAT scores for a random sample of 10 undergraduate students from a randomly selected college and another sample of undergraduates from a university center.

Institution	SAT Scores
College	1330, 1330, 1340, 1370, 1380, 1470, 1340, 1450, 1450, 1360
University	1180, 1160, 1140, 1390, 1380, 1320, 1150, 1240, 1380, 1150

9. Do the students from the college have higher SAT scores than those from the university center?

10. Create an appropriate visualization of these data to accompany your result from the above analysis.

CHAPTER 8

TESTING DIFFERENCES AMONG MULTIPLE SAMPLES

We often are interested in testing the effect of a factor over a variety of levels. If, for instance, we're testing the effect of nutrient addition on an organism's growth rate, we might have a treatment without fertilizer (called a "control"; see section 6.4), as well as levels with medium and high fertilizer. When we have more than two sample groups we may not just do multiple t-tests. Why not? The bottom line is that conducting multiple tests, like many t-tests, will increase our chance of making a Type I error. You may recall from section 6.7 that this happens when we erroneously reject a true H_0. The use of a single test, like an analysis of variance (ANOVA), reduces our likelihood of making this mistake.

8.1 SAMPLES ARE NORMALLY DISTRIBUTED

When you have more than two samples and each sample is normally distributed (see section 4.4) then, using a parametric test (see section 4.4), you are able to ask the question as to whether the samples come from the same population or not.

Let's assume we have data on the mass of young fish that were grown using three different food supplemental rates. When we have more than two samples and the data are normally distributed, then we are able to use an analysis of variance (ANOVA) to test whether there was an effect of food treatment on fish growth, measured as the mass of fish grown over some length of time (here, one month). One challenging part of this test is getting the data into the proper format. To do this I have simply entered the mass of each fish grown under each treatment level into its own variable. Then it's simple enough to gather them into a dataframe.

```
> Low = c(52.3, 48.0, 39.3, 50.8, 53.3, 45.1)
> Med = c(50.4, 53.8, 53.4, 58.1, 56.7, 61.2)
> High = c(66.3, 59.9, 59.9, 61.3, 58.3, 59.4)
> my.fish.dat = data.frame(Low,Med,High)
> my.fish.dat = stack(my.fish.dat)
```

We could have done this using Excel and had each set of data in its own column. However, there aren't a lot of data in this case so I just entered the data in a script file and created a data frame called my.fish.dat, and then "stacked" the data. When we have more than two samples, we need them to be stacked to graph them and to do our analyses (see section 3.3). The default names for the variables after stacking are "values" and "ind." I like to change these so that they are more meaningful using the names() function:

```
> names(my.fish.dat) = c("Mass","Trmt")
```

Now I'm ready to visualize these data. Probably the simplest and best way to see a comparison of the distributions is to use a side-by-side boxplot (see Figure 8.1). I have the data in separate variables (Low, Med, and High) and as separate variables. The boxplot for these data can be made easily with those (see Figure 8.1).

```
> boxplot(Low,Med,High, names = c("Low","Medium","High"),
+         xlab = "Feeding Rate", ylab = "Fish Mass (g)",
+         cex.lab = 1.5)
```

Alternatively, if the data are only in a stacked dataframe then R will order the boxplots alphabetically. This usually isn't what we want. To adjust the order of the boxes we need to tell R the order of these factor levels within the my.fish.dat dataframe. Below is how we do that using the factor() function. With this function I specify the order that I want the factor levels to appear. The graph looks just like Figure 8.1 and, therefore, is not reproduced here.

```
> my.fish.dat$Trmt = factor(my.fish.dat$Trmt,
+         levels = c("Low","Med","High"))
```

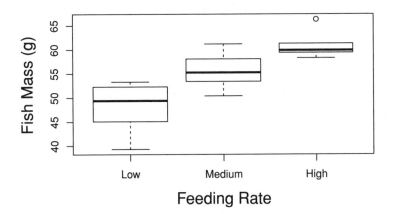

Figure 8.1: The mass of fish grown under three food treatments. The samples are significantly different (F = 202; df = 2, 15; p < 0.001).

```
> boxplot(my.fish.dat$Mass ~ my.fish.dat$Trmt,
+        names = c("Low","Medium","High"),
+        xlab = "Feeding Rate", ylab = "Fish Mass (g)",
+        cex.lab = 1.5)
```

These data appear to be normally distributed. We can complete this test like this for brevity:

```
> shapiro.test(Low)$p.value
```

```
[1] 0.4883008
```

```
> shapiro.test(Med)$p.value
```

```
[1] 0.9592391
```

```
> shapiro.test(High)$p.value
```

```
[1] 0.05810509
```

and see that they do appear to be normally distributed (p-values > 0.05; see section 4.4 on page 56)

We can now perform the analysis of variance with the following call:

```
> my.aov = aov(my.fish.dat$Mass~my.fish.dat$Trmt)
```

and look at the output, stored in the variable my.aov with the following call:

```
> summary(my.aov)
```

```
                   Df Sum Sq Mean Sq F value   Pr(>F)
my.fish.dat$Trmt   2  490.1  245.03   14.55 0.000308 ***
Residuals          15 252.7   16.84
---
Signif. codes:  0 '***' 0.001 '**' 0.01 '*' 0.05 '.' 0.1 ' ' 1
```

The summary provided above is called an "analysis of variance table." At first we usually are most interested in whether the samples are statistically different. We see this, like usual, by inspecting the p-value. This is found on the right side of this table under the Pr(>F) heading where we see p = 0.0003. Therefore, we can conclude that the masses of fish grown under the different feeding treatments are statistically different (F = 14.55; df = 2, 15; p < 0.001). Notice that this is correct but rather empty since I'm not able to say much about which groups differ from which and in what way. We need to do a little more work to interpret these more complicated datasets and results.

INTERPRETING RESULTS FROM A ONE-WAY ANOVA

We have seen that feeding rates affect fish masses (Figure 8.1). Increasing feeding from low to medium seems important, but increasing from medium to high rates seems to have a smaller effect. To test for differences among levels in an ANOVA we perform what's called a "*post hoc*" test (i.e., after the fact test). I often use Tukey's Honest Significant Differences test (the TukeyHSD() function in R). It's a good *post hoc* test if the variances of the different samples are similar, which is an assumption of the analysis of variance. Other tests could be used instead (but choose only one!). To implement this test we send the variable my.aov to the TukeyHSD() function:

```
> TukeyHSD(my.aov)
```

```
  Tukey multiple comparisons of means
    95% family-wise confidence level

Fit: aov(formula = my.fish.dat$Mass ~ my.fish.dat$Trmt)
```

```
$`my.fish.dat$Trmt`
                diff       lwr       upr     p adj
Med-Low   7.466667   1.311785  13.62155 0.0170662
High-Low 12.716667   6.561785  18.87155 0.0002185
High-Med  5.250000  -0.904882  11.40488 0.1008030
```

The interpretation of the output from this test is a bit challenging. The function provides a table with each pairwise comparison on the left (e.g., low-high) with a difference between the means (diff). It also provides lower (lwr) and upper (upr) bounds. If those bounds include 0, then the two samples are not statistically different. We also get an associated p-value (p adj) for each comparison in the right column. In this example two of the three comparisons are significantly different. However, we see that the medium and high treatment groups are not statistically different. As we can see, not all samples need to be significantly different from each other for the overall ANOVA to be significant.

We saw these data graphed using boxplots in Figure 8.1. This is a good visualization to see the differences among the samples. However, your instructor, and many publications, expect these data to be presented as barplots with error bars and possibly more information. This is a more advanced topic and is discussed in detail later in this book (see Figure 11.2 on page 167).

8.2 ONE-WAY TEST FOR NON-PARAMETRIC DATA

If the samples are not normally distributed then we can conduct a Kruskal-Wallis test. Keep in mind that this test does not assume samples are normally distributed but it does assume that the samples come from the same population (the standard H_0). Therefore, all the samples are assumed to come from the same shaped distribution. If your sense is that the distributions of the samples are similar then you may proceed with this test. Below are data for the lengths of tibia in three species of small mammals:

```
> A = c(58.4, 55.9, 53.8, 53.8, 53.5, 58.7, 52.9,
+       57.5, 58.5, 53.3)
> B = c(70.9, 74.9, 70.6, 71.9, 71.0, 72.2, 74.5,
+       62.0, 69.6, 60.2)
> C = c(70.3, 72.2, 71.6, 71.5, 73.3, 72.2, 77.7,
+       77.5, 73.3, 71.7)
```

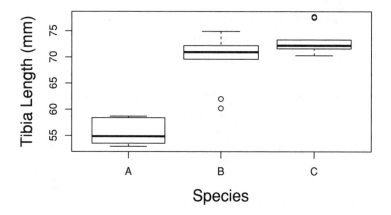

Figure 8.2: Boxplot of the distribution of tibia lengths for three species. The three samples were statistically different (K= 9, $\chi^2 = 20.6$, p < 0.001).

Now let's create an appropriate graph to view the distributions of our three samples and compare them side-by-side (see Figure 8.2).

```
> boxplot(A,B,C,names = c("A","B","C"),
+           ylab = "Tibia Length (mm)",
+           xlab = "Species",cex.lab = 1.5)
```

Looking at these distributions it's hard to tell whether they're normally distributed. As described earlier (see section 4.4 on page 56), let's test our samples for normality using the Shapiro-Wilk test and see what it suggests:

```
> shapiro.test(A)$p.value # normal?

[1] 0.03191903

> shapiro.test(B)$p.value # normal?

[1] 0.02546059

> shapiro.test(C)$p.value # normal?
```

[1] 0.01866955

Note that we test each sample individually. Do not combine the data into one set and then test it. The above tests suggest that our samples do not appear to be normally distributed. Therefore, we will proceed with the non-parametric Kruskal-Wallis test for comparing these samples.

To conduct this test we need to have the data stacked, as we did for the analysis of variance. For brevity I've done this below and accepted the new column names provided by R.

```
> my.dat = data.frame(A,B,C)
> my.dat = stack(my.dat)
> kruskal.test(values ~ ind, data = my.dat)

        Kruskal-Wallis rank sum test

data:  values by ind
Kruskal-Wallis chi-squared = 20.6314, df = 2, p-value
= 3.311e-05
```

From this analysis we find that our data do not support the null hypothesis that the samples come from the same population. Instead, we found that tibia lengths for the three species measured were statistically different ($\chi^2 = 20.6$, df = 2, p < 0.001).

If you need to conduct a *post hoc* test for the Kruskal-Wallis test you might do this using the `kruskalmc()` function found in the **pgirmess** package.

8.3 TWO-WAY ANALYSIS OF VARIANCE

In this section we discuss a more complicated design and will introduce only the test that assumes the data are normally distributed. What do we mean by a "two-way" (or "two-factor") test? Sometimes we are interested in testing whether two factors (perhaps watering rates and nutrients) interactively affect our system, such as plant growth. This seems really complicated. You might be tempted to approach this two-factor problem as two separate experiments. We could, for instance, test whether different water levels affect plant growth. We might find that plants grow more with more water. Likewise, we might want to test the effect of fertilizer levels on plant growth. Again, we might find that plants grow more with more fertilizer. Why consider a two-factor test? Here are two important reasons:

Table 8.1: Design for 2×2 factorial experiment to determine the effects and interaction of water and nutrient levels.

Number of Plants	Water Level	Nutrient Level
8	Low	Low
8	Low	High
8	High	Low
8	High	High

1. Experiments are expensive and we want to save money.
2. We want to know if our factors *interact*.

Saving money is good. But the second reason is the clincher. It turns out that plants often will respond to fertilizer only if they have enough water. If they have low levels of water available then fertilizing them can actually kill them! So, the growth rate of our plants can depend not only on water and fertilizer levels, but also on the *interaction* between water and fertilizer. Just to restate this problem: if we test two factors in separate experiments we would be unable to discover that the two factors might interact. A multi-factor design is more complicated and harder to analyze but allows us to ask more interesting questions in biology. If we shy away from this complexity and avoid complicated statistical tests then we very well can miss exciting aspects of biological systems. Designs can actually be much more complicated than this, and this complexity is where we are in biology! The simple questions have been asked and answered and, therefore, are no longer fundable or publishable. So, buckle up!

The experiment I'm going to show you is called a 2×2 factorial design (we say "two by two factorial design") (see Table 8.1). We might have a third factor with three levels (testing how water and nutrient levels affect three different species). We would then have a $2 \times 2 \times 3$ factorial design. If each treatment group had eight replicates how many individual sample units do we need? We just need to multiply:

Design $= 2 \times 2 \times 3$ factorial experiment with 8 replicates per treatment
2 water levels $\times$ 2 nutrient levels $\times$ 3 species $\times$ 8 reps $= 96$ plants.

The design shown in Table 8.1 is a 2×2 factorial design and, as shown, requires 32 plants ($2 \times 2 \times 8$ replicate plants per treatment). Here are the data for dried plant masses in grams.

```
> LW.LN = c(3.84,4.47,4.45,4.17,5.41,3.82,3.83,4.67)
> LW.HN = c(8.84,6.54,7.60,7.22,7.63,9.24,7.89,8.17)
> HW.LN = c(7.57,8.67,9.13,10.02,8.74,8.70,10.62,8.23)
> HW.HN = c(16.42,14.45,15.48,15.72,17.01,15.53,16.30,15.58)
```

The variables are each named with four letters (e.g., LW.LN). The first two letters represent the water levels (low water (LW) or high water (HW)) and the last two letters represent the nutrient levels (either low nutrient (LN) or high nutrient (HN)). We're going to now name these treatment levels appropriately using the function rep(), which stands for "repeat."

```
> water = rep(c("LW","HW"), each = 16)
> nutr = rep(c("LN","HN"), each = 8, 2)
> plant.mass = c(LW.LN,LW.HN,HW.LN,HW.HN) # combine the data
```

We can see what water and nutr now look like by printing them to the screen:

```
> water

 [1] "LW" "LW" "LW" "LW" "LW" "LW" "LW" "LW" "LW" "LW" "LW"
[12] "LW" "LW" "LW" "LW" "LW" "HW" "HW" "HW" "HW" "HW" "HW"
[23] "HW" "HW" "HW" "HW" "HW" "HW" "HW" "HW" "HW" "HW"
```

```
> nutr

 [1] "LN" "LN" "LN" "LN" "LN" "LN" "LN" "LN" "HN" "HN" "HN"
[12] "HN" "HN" "HN" "HN" "HN" "LN" "LN" "LN" "LN" "LN" "LN"
[23] "LN" "LN" "HN" "HN" "HN" "HN" "HN" "HN" "HN" "HN"
```

The variable plant.mass contains our data. Now we can create the dataframe, using the function data.frame(), and store the data in a variable called P.dat. We can send this function our three arrays of data from above and it will assemble everything into one structure.

```
> P.dat = data.frame(water,nutr,plant.mass)
```

We can view the beginning of this structure with the head() function like this:

```
> head(P.dat) # view the first 6 rows

  water nutr plant.mass
1    LW   LN       3.84
2    LW   LN       4.47
3    LW   LN       4.45
4    LW   LN       4.17
5    LW   LN       5.41
6    LW   LN       3.82
```

We see that the two factors (**water** and **nutr**) are at the top of the columns. Our data are in the **plant.mass** column. If we look at the whole dataframe we'd see **HW** and **HN**, as well. We can now create a boxplot of the data using the formula approach (see Figure 8.3). Note that we have yet to conduct a hypothesis test.

```
> boxplot(P.dat$plant.mass ~ P.dat$water * P.dat$nutr,
+          ylab = "Biomass (g)", xlab = "Treatment",
+          las = 1, cex.lab = 1.5)
```

Now that we've seen what our results look like it's time to begin our analysis. The H_0 is that the samples have all been drawn from a single population (that their means and variances are all the same). That doesn't look likely but we need some quantitative support for our rejection of the H_0.

The analysis of variance is a parametric test and assumes that the data are normally distributed. A lot of work has gone into assessing how robust ANOVA is to deviations from normality and, it turns out, the evidence suggests ANOVA is very robust. We will still test for normality and have to do this for each sample separately. Pulling out just our samples from a dataframe requires subsetting our data. Fortunately, we've done subsetting before (see section 3.4 on page 40). Here's another way we can subset our data. Let's imagine that we want just the plant masses for the treatment when water was low and nutrient levels were low. This can be done like this:

```
> P.dat$plant.mass[P.dat$water == "LW" &
+                  P.dat$nutr == "LN"]

[1] 3.84 4.47 4.45 4.17 5.41 3.82 3.83 4.67
```

What we're asking R to do is to print to the screen the **plant.mass** data from the dataframe **P.dat**. We don't want everything. Instead, we only

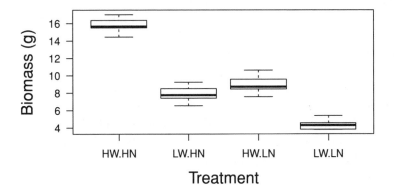

Figure 8.3: A quick visualization of our samples. We see the samples all appear relatively normally distributed. Note the naming scheme used where the first two letters represent the water treatment level and the latter two letters represent the nutrient levels.

want the `plant.mass` when the water level is equivalent to `LW` (that's the `==` combination), which is low water **and** (that's the & symbol) when the nutrients are low (`LN`).

Now that we know how to subset our dataframe we can send those data as samples to the `shapiro.test()` function to test for normality. I have asked R to provide just the p-values for this test by adding the "`$p.value`" to the function call.

```
> shapiro.test(P.dat$plant.mass[water == "LW" &
+                               nutr == "LN"])$p.value

[1] 0.1734972

> shapiro.test(P.dat$plant.mass[water == "LW" &
+                               nutr == "HN"])$p.value

[1] 0.9509188

> shapiro.test(P.dat$plant.mass[water == "HW" &
+                               nutr == "LN"])$p.value
```

```
[1] 0.6769018
```

```
> shapiro.test(P.dat$plant.mass[water == "HW" &
+                          nutr == "HN"])$p.value
```

```
[1] 0.744246
```

As you might notice, pulling out the individual samples is a little tough. I could have done this more simply with just testing each individual variable that we had before joining them into a dataframe (e.g., LL) but it's important to know what to do if you have the data stacked. So, the data appear to be normally distributed. Knowing that the data appear normal, having tested this using the Shapiro-Wilk test, we may proceed with the ANOVA hypothesis test.

I can now run the ANOVA with the following call, using the aov() function and storing my result in a variable called my.aov2:

```
> my.aov2 = aov(P.dat$plant.mass ~
+               P.dat$water * P.dat$nutr)
```

The model that we want to fit is sent to the aov() function as y ~ x_1 * x_2. The multiplication on the right-hand side tells R that we want the main effects and the interaction term.

R doesn't report anything because the output is, instead, stored in the variable my.aov2. We should use this approach of saving the output from the aov() function so we can then get and format the output in more useful ways. You can see the results by typing my.aov2 at the command prompt. You can have R format the output for you by sending the variable my.aov2 to the summary() function. R will return the ANOVA table.

```
> summary(my.aov2)
```

	Df	Sum Sq	Mean Sq	F value	Pr(>F)
P.dat$water	1	314.88	314.88	488.69	< 2e-16
P.dat$nutr	1	216.74	216.74	336.37	< 2e-16
P.dat$water:P.dat$nutr	1	21.68	21.68	33.65	3.13e-06
Residuals	28	18.04	0.64		

| P.dat$water | *** |
| P.dat$nutr | *** |

```
P.dat$water:P.dat$nutr ***
Residuals
---
Signif. codes:  0 '***' 0.001 '**' 0.01 '*' 0.05 '.' 0.1 ' ' 1
```

The output from R is a bit messy, and awkwardly wrapped here, but the information is important for us to understand what happened in our experiment. I've reformatted the output into Table 8.2. In the next section we'll discuss in greater detail what this means.

Table 8.2: ANOVA table for the effect of water and nutrient levels on plant biomass.

	Df	Sum Sq	Mean Sq	F value	Pr($>$F)
P.dat$water	1	314.88	314.88	488.69	0.0000
P.dat$nutr	1	216.74	216.74	336.37	0.0000
P.dat$water:P.dat$nutr	1	21.68	21.68	33.65	0.0000
Residuals	28	18.04	0.64		

INTERPRETING THE ANOVA TABLE

As with our other statistical tests we are quite interested in the p-value so our eyes naturally go to see what happened. That's the far right column in Table 8.2. We see the p-value is in the column labeled Pr(>F). You notice that they're all zero! Well, they are **not** zero; they are just small. We should report them as p < 0.001 in a laboratory report (you might revisit our discussion of p-values in section 6.6 on 94).

There are three "effects" in the table, one for each of our two factors, called "main effects," and a third for the "interaction effect" between these two factors. The first main effect is for the effect of water on plant mass. This line gives us the information we need to interpret the effect of our watering levels on plant biomass. We go across this row and see that the degrees of freedom (Df) for this term is 1. This tells us the number of levels, which is one more than Df (Df $= 2 - 1 = 1$). So, there must have been two levels. The mean square error term (Mean Sq) is the sums of squares (Sum Sq) divided by the degrees of freedom (Df). The F value is the Mean Sq term divided by the residuals Mean Sq term in the Residuals row (0.64). The F statistic is used to determine a p-value which, in R, is labeled "Pr(>F)." In general, the bigger the F value the smaller our p-value will be. It's our job to compare the p-values against our pre-determined α, which we usually set at 0.05, to determine statistical significance.

The second row is the main effect for the nutrient treatment. The last row represents the "interaction term" for the water by nutr treatment effect. Some researchers consider the interaction term to be more important than the main effects and even use a different (usually higher) critical p-value (α) to determine the importance of this effect. Check with your laboratory instructor or mentor in how they want you to interpret "higher-order" effects.

Our results suggest strong main effects for both water level and nutrient level on plant biomass. In general we see that plants with high nutrient levels (the two boxplots on the left in Figure 8.3) are larger, on average, than plants with low nutrient levels (the two boxplots on the right in Figure 8.3). We also see that plants with higher water levels (HW) tend to be larger than plants with lower water levels (LW). Does that make sense? It's always important, and sometimes tricky in these tests, to know what the results should look like. Be careful to not just plug and chug statistical tests.

The interaction term can be visualized by using the interaction.plot() function (see Figure 8.4 for a description of what this plot shows). This is an important graph for visually inspecting statistical significance. If the interaction term is not significant these lines will be approximately parallel. In this example the lines are not parallel, which is consistent with the significant water by nutrient interaction.

```
> interaction.plot(P.dat$water, trace.label = "nutr",
+       trace.factor = P.dat$nutr, P.dat$plant.mass)
```

To determine whether individual samples differ from each other we need to invoke the *post hoc* comparison (discussed above in section 8.1 on page 120). Here's how to do this with our data:

```
> TukeyHSD(my.aov2)

  Tukey multiple comparisons of means
    95% family-wise confidence level

Fit: aov(formula = P.dat$plant.mass ~ P.dat$water * P.dat$nutr)

$`P.dat$water`
          diff      lwr       upr p adj
LW-HW -6.27375 -6.855088 -5.692412     0
```

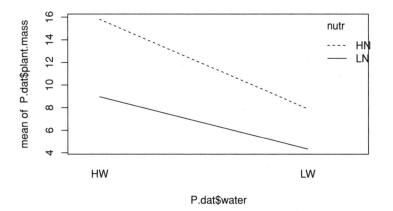

Figure 8.4: An interaction plot for the plant biomass data. Notice that if the lines connecting the means of the nutrient factor (high and low nutrients in this example) are parallel then this suggests there is no "statistical interaction." If the lines are not parallel then it suggests there is a significant interaction. Since we know, from our ANOVA table (see Table 8.2), this interaction is statistically significant (F = 33.65; df = 1, 28; p < 0.001), the lines should not be parallel, which appears to agree with the graph.

```
$`P.dat$nutr`
        diff       lwr       upr p adj
LN-HN -5.205 -5.786338 -4.623662     0

$`P.dat$water:P.dat$nutr`
                  diff          lwr         upr       p adj
LW:HN-HW:HN   -7.92000  -9.01582151  -6.824178 0.0000000
HW:LN-HW:HN   -6.85125  -7.94707151  -5.755428 0.0000000
LW:LN-HW:HN  -11.47875 -12.57457151 -10.382928 0.0000000
HW:LN-LW:HN    1.06875  -0.02707151   2.164572 0.0579272
LW:LN-LW:HN   -3.55875  -4.65457151  -2.462928 0.0000000
LW:LN-HW:LN   -4.62750  -5.72332151  -3.531678 0.0000000
```

The output from this test at first appears to be quite confusing. There's a lot of information here. We see the results for the two main effects (water and nutr) individually, followed by the interaction comparison (below the $'P.dat$water:P.dat$nutr'). For the interaction comparison each of the six rows represents a statistical, pair-wise test between two treatments. The first row, for instance, is for the LW:HN-HW:HN comparison. What's that? This is a comparison between the mean for plants that receive low water and high nutrients (LW:HN) against the mean for those grown under high water and high nutrients (HW:HN). The last column shows the result of the statistical test ("p adj"), which is the p-value. You can see that they are all statistically significant ($p < 0.001$) except for one comparison which has $p = 0.0579$. Can you find that comparison in Figure 8.4?

This *post hoc* test on the interaction term only should be done if the interaction term is statistically significant. Once we do this *post hoc* test we can interpret which of the samples differ from each other.

VISUALIZING THE RESULTS OF A TWO-WAY ANOVA TEST

Interpreting a two-way analysis of variance, as we have seen, can be tricky. The interaction plot (Figure 8.4) helps us to understand the results from our experiment but falls short of a professional-looking graph. Instead, the standard practice is to use a barplot which groups together one of the factors. Later, in section 11.2 on page 164, we'll see how to add error bars and letters to this plot to make it publication quality.

If all you need at this point is a barplot of the means then let's see how to do that. To build this graph we need to have our data in a "matrix" which is a two-dimensional array. This way we can group the bars by the factor of our choice (see Figure 5.9 on page 80).

Here's how to get the means from our data and store the results in a matrix. The tapply() function performs a cross-tabulation on the data and applies the mean() function to the data by groups.

```
> M = tapply(P.dat$plant.mass,
+            list(P.dat$water, P.dat$nutr),
+            mean)
```

We can now use matrix M to create our barplot, as is shown in Figure 8.5 (also see the barplot section on page 72).

```
> barplot(M, beside = TRUE, ylim = c(0,20),
+    xlab = "Nutrient Level", ylab = "Mass (g)",
```

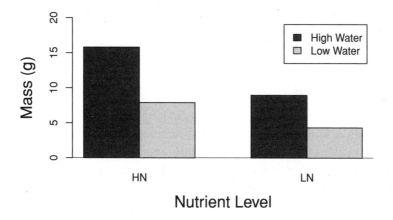

Figure 8.5: A barplot of the two-way analysis of variance.

```
+    legend = c("High Water","Low Water"),
+    cex.lab = 1.5)
> abline(h=0)
```

If we want to switch the factor on the x-axis and the trace factor (water, which shows up in the legend) then we just reverse the order of the factors in the list in the `tapply()` function and change the legend text to be high and low nutrients.

8.4 PROBLEMS

(For questions 1–5). Below are data for the bacterial counts found in milk from different dairies. The columns represent different farms and the rows are the samples (six from each of the five farms). Units are in thousands of colony forming units per milliliter (cfu ml^{-1}).

Farm 1	Farm 2	Farm 3	Farm 4	Farm 5
24	14	11	7	19
15	7	9	7	24
21	12	7	4	19
27	17	13	7	15
33	14	12	12	10
23	16	18	18	20

1. Test whether the data are normally distributed.
2. Stack the data in a dataframe called `milk.dat2`.
3. Is there a difference in bacterial counts among the farms?
4. If so, describe in words which are different from which?
5. Create a barplot of the bacterial counts as a function of the farm.

(For questions 6–10). Table 8.3 on page 135 contains data on the lengths of twenty stickleback fish grown under four different conditions (cold/warm water and high/low pH tanks). The lengths are in centimeters. Use these to answer the following questions.

6. Enter these data into a spreadsheet as you see them. Save them in the `.csv` format. This is the standard stacked form so that any statistics package can handle them. Read the spreadsheet into an R dataframe called `my.sticks`. Verify that they are entered correctly in R by typing `my.sticks` at the console.
7. Test whether the samples are normally distributed.
8. Perform a two-way analysis of variance on lengths. What does this suggest?
9. Provide two side-by-side barplots of the main effects (temperature and pH).
10. Provide a single barplot of the interaction term for this analysis. What do you conclude about the effects of temperature, pH, and the interactive effects of these two factors?

Table 8.3: Design for 2×2 factorial experiment to determine the effects and interaction of water and nutrient levels.

Length	Temp	pH	Length	Temp	pH
C	H	4.6	W	H	4.9
C	H	4.6	W	H	4.8
C	H	4.3	W	H	4.5
C	H	4.5	W	H	4.8
C	H	4.7	W	H	4.8
C	L	3.4	W	L	5.8
C	L	3.0	W	L	5.6
C	L	3.2	W	L	5.8
C	L	2.9	W	L	6.0
C	L	3.2	W	L	5.9

CHAPTER 9

HYPOTHESIS TESTS: LINEAR RELATIONSHIPS

We are often interested in asking whether two variables are related. For instance, we might ask whether the mass of dogs is related to the length of dogs. In this kind of test the data are generally continuous and paired together somehow (a dog has a mass and a length). What? Two measurements from the same experimental unit? It's OK because we use each dog (name and mass) as a single datum (point!). We do, however, need to have independence between our data points. We wouldn't want to test this relationship between dog lengths and masses using only dogs from a single litter (they wouldn't be independent), unless we're asking something about this single litter, which is doubtful we'd want to do.

There are two broad types of tests we might perform on such data, assuming we're investigating whether there is a linear relationship between two variables. These tests are correlation and regression analyses. We perform a correlation analysis when we're interested in determining if there is a linear relationship between two normally distributed variables and a regression analysis when we're interested in testing whether one normally distributed variable is dependent on the other variable. Strangely, the correlation test assumes a linear relationship but we never add a best-fit line to the data! This can cause a lot of confusion as to whether you should add a line to a scatterplot.

Should I add a line to a scatterplot?

Correlation. No. There is NEVER a best-fit line added to a correlation plot. There is no causality (or "dependence") implied. Either variable can be plotted on the y-axis. A correlation analysis results in a correlation coefficient (**r**) and a p-value.

Regression. Maybe. The dependent variable (response variable) is placed only on the y-axis. The independent variable (predictor variable) goes on the x-axis. You may add a line to regression data only if the relationship is statistically significant. Provide the reader with the statistical output (F; df_1, df_2; p-value). You also might report the adjusted R^2 value (check with an instructor as to what is wanted). If the relationship is statistically significant add the line to the graph and report the equation (e.g., $y = 2.7x + 14$).

9.1 CORRELATION

As discussed above, a correlation analysis is a test that investigates whether two variables are linearly related. Let's assume we are interested in the relationship between the number of bars and the number of churches in a variety of towns across a state (e.g., I'm in New York State right now). Here are my data for 10 towns of different sizes:

```
> churches = c(29, 44, 45, 46, 53, 53, 54, 57, 70, 90)
> bars = c(20, 22, 36, 37, 38, 41, 60, 72, 76, 99)
> plot(churches,bars, cex = 1.5, pch = 19,
+      xlab = "Number of Churches",ylab = "Number of Bars",
+      xlim = c(0,100), ylim = c(0,100),cex.lab = 1.5)
```

I have graphed them in Figure 9.1. There seems to be a positive, linear relationship. We might wonder if we should put a best-fit line through data points. Is the number of bars (y-axis variable) *dependent* on the number of churches (x-axis variable) in different towns?

The proper way to approach these data is to first graph them, as we have done. You might ask yourself if they're graphed correctly. I've placed the number of bars on the y-axis and the number of churches on the x-axis. Could

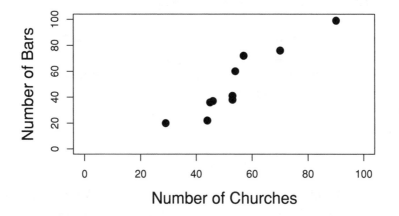

Figure 9.1: The relationship between number of churches and the number of bars in 10 towns of different sizes. There appears to be a strong, positive relationship.

I switch them around? Actually, I think I can. It doesn't seem to matter what's on the y-axis. This is because the y variable, in this case the number of bars, does not seem to depend on how many churches there are in a town. Instead, both these variables depend on the number of people living in these towns.

It's time to perform the correlation analysis. The correlation analysis is a parametric test and assumes both variables are normally distributed. Below I test this but report only the p-values.

```
> shapiro.test(churches)$p.value
```

```
[1] 0.2950524
```

```
> shapiro.test(bars)$p.value
```

```
[1] 0.3281046
```

Now we may proceed with the correlation test using the `cor.test()` function. Note that the H_0 is that there is no correlation.

```
> cor.test(bars,churches) # Are these correlated?

        Pearson's product-moment correlation

data:  bars and churches
t = 7.0538, df = 8, p-value = 0.0001068
alternative hypothesis: true correlation is not equal to 0
95 percent confidence interval:
 0.7183203 0.9832072
sample estimates:
      cor
0.9281631
```

R provides us a lot of information from this analysis. We see the type of test (Pearson's product moment correlation test), the test results, and the correlation coefficient. With this analysis we conclude that there is a highly significant, positive correlation between the number of bars and the number of churches in 10 towns in New York State ($r = 0.928$, $t = 7.05$, $df = 8$, $p < 0.001$).

CORRELATION WITH MANY VARIABLES

You can perform a correlation analysis on many variables at one time. First, it's great to use the exploratory graphing function **pairs()** to visualize all of the relationships between all our variables. We can use this graph to find relationships that might be of potential interest.

Back in section 4.7 on page 65 I asked you to go to http://data.worldbank. org/indicator/SP.POP.TOTL and download the Excel spreadsheet data for populations by country. If you haven't please do so, save the file in the .csv file format, and read the data into a dataframe called world.pop. There are population estimates for almost all countries on Earth from 1960 to 2012. I have saved these into my working directory and will read them into the dataframe as follows:

```
> world.pop = read.csv("worldpop.csv") # name I gave the file
```

I'm interested in asking the question whether populations for different countries changed in similar ways. The dataset, however, is really big. I'm going to just investigate this for three years. I need the names of the columns in order to make the call to make this graph. Here are what the first size column headers look like. You can look at the rest without the **head()** function.

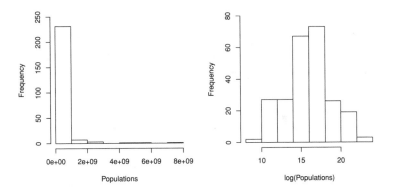

Figure 9.2: Two histograms of the world population estimated for all countries in 2012, reported by the World Bank (see `http://data.worldbank.org/indicator/SP.POP.TOTL`.

```
> head(names(world.pop))

[1] "Country.Name" "Country.Code" "X1960"
[4] "X1961"        "X1962"        "X1963"
```

I'm going to look at just one of the years to see what the distribution looks like:

```
> par(mfrow = c(1,2))
> hist(world.pop$X2012,main = "", ylim = c(0,250),
+      xlab = "Populations")
> hist(log(world.pop$X2012), main = "", ylim = c(0,80),
+      xlab = "log(Populations)")
> par(mfrow = c(1,1)) # restore graphics window
```

As you can see the data for 2012 are skewed to the right (left graph, Figure 9.2). I can transform these with the `log()` function and, as you can see, the data are close to normally distributed on this log scale (right graph, Figure 9.2). I now want to see how different years might be related between the years 1960, 1985, and 2012 (over a 50-year period). To do this I can use the `pairs()` function but I need the data in a matrix. Here's how to gather up the three different years into a dataframe. Note that I can log all the data at once when they are in a dataframe.

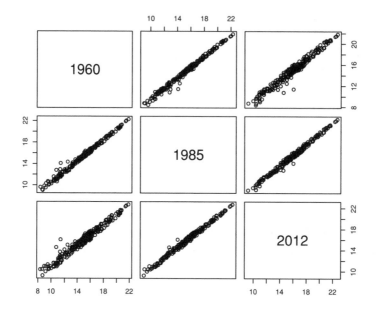

Figure 9.3: A comparison of the logged population data for 1960, 1985, and 2012. We can see that the population sizes for countries are strongly related. This means that countries that were small in 1960 remained small even in 2012, and likewise for large countries. Data from `http://data.worldbank.org/indicator/SP.POP.TOTL`.

```
> pop = data.frame(world.pop$X1960,world.pop$X1985,
+                  world.pop$X2012)
> pop = log(pop) # log all the data
> names(pop) = c("1960","1985","2012") # give them names
> pop = pop[complete.cases(pop),] # remove missing values
> pairs(pop) # Note that the pop data are logged
```

This type of figure is great for data exploration, allowing us to easily visualize whether different variables might be positively or negatively related, or not related at all (see Figure 9.3).

We can evaluate these relationships quantitatively by examining the "correlation matrix" of these data using the `cor()` function with the following code:

```
> cor(pop)

          1960       1985       2012
1960 1.0000000 0.9934600 0.9787352
1985 0.9934600 1.0000000 0.9939553
2012 0.9787352 0.9939553 1.0000000
```

The command `cor(pop)` provides you all the pair-wise correlation coefficients (`r`) for the logged data. Notice that on the main diagonal $r = 1.0$. This is because all the variables are perfectly correlated with themselves. At this point we have not conducted a hypothesis test. It's always good to keep the number of statistical tests to a minimum to avoid making statistical "errors" (see section 6.7) resulting from a "fishing expedition" (using every known statistical test in hopes of finding *something* significant). If, after careful consideration, you want the p-value for a particular pairing then run a test:

```
> names(pop)

[1] "1960" "1985" "2012"

> cor.test(pop$"1960",pop$"2012")

        Pearson's product-moment correlation

data:  pop$"1960" and pop$"2012"
t = 73.7631, df = 239, p-value < 2.2e-16
alternative hypothesis: true correlation is not equal to 0
95 percent confidence interval:
 0.9726675 0.9834671
sample estimates:
      cor
0.9787352
```

For this analysis we conclude that the population sizes for the countries are positively correlated (r =0.979, t = 73.76, df = 239, p < 0.001).

9.2 LINEAR REGRESSION

The linear regression is used when you know that **y** depends on **x** in a linear fashion (or that relationship has been made linear through a transformation, like logging the data). It assumes the y-variable is normally distributed (see section 4.4 on page 56).

Some people think the goal of a regression is to just add a line on a scatterplot because the data points look naked. However, the meaning of a best-fit relationship is much more interesting. We often have worked hard and spent time and money to get each value in our dataset. We consider each value as giving us insider information as to how the world works.

We usually perform a regression analysis because we are interested in testing a hypothesis that relates one variable directly to another variable through some dependency. We might, for instance, be asking a question about whether something increases, or just changes, as a function of time. This would suggest we're actually interested in knowing the slope of the relationship, which often represents a rate. We also are often interested in where the line crosses the y-axis, called the "y-intercept." Our methods here will allow us to formalize both our estimates of these parameters (slope and intercept) and also provide us the estimates of error for these parameters.

Let's consider the example of the relationship between blood alcohol concentration (BAC) and the number of drinks consumed for a 160 lb male. Here are the estimated data (from `http://www.faslink.org/bal.htm`):

```
> drinks = 1:10
> BAC = c(0.02, 0.05, 0.07, 0.09, 0.12, 0.14, 0.16, 0.19,
+           0.21, 0.23)
```

The first thing we should consider is that there might be a relationship between these variables. Does having more drinks influence BAC? Probably! If we graph these, which of the variables goes on the y-axis? If BAC depends on the number of drinks someone has then it should go on the y-axis. But can we switch the axes? Can the number of drinks just as easily go on the y-axis? Let's graph both and look.

```
> par(mfrow = c(1,2))
> plot(drinks,BAC, pch= 16, cex.lab = 1.5, xlim = c(0, 11),
+       ylim = c(0,0.3), xlab = "Number of Drinks")
> plot(BAC,drinks, pch= 16, cex.lab = 1.5, ylim = c(0, 11),
+       xlim = c(0,0.3), ylab = "Number of Drinks")
> par(mfrow = c(1,1))
```

The only plausible graph is the left panel. Blood alcohol content depends on how many drinks have been consumed, not the other way around. So we

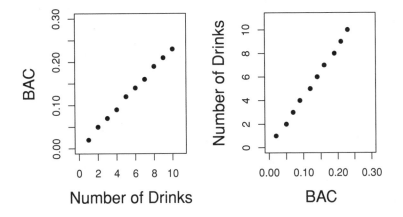

Figure 9.4: The left panel shows blood alcohol content (BAC) graphed as a function of the number of drinks for males of approximately 160 pounds taken in one hour. The right panel switches these axes. Are these both plausible ways to graph these data? Data from http://www.faslink.org/bal.htm.

now have a relationship that exhibits a dependency. And the data appear to be linearly related so we can begin assessing this relationship using linear regression.

Linear regression analysis is a parametric test and assumes that the y-axis data are normally distributed. Let's test this:

```
> shapiro.test(BAC)$p.value
```

```
[1] 0.8790782
```

The p-value is greater than 0.05, which suggests we not reject the H_0 that these data are normally distributed. Therefore, we may proceed with the regression analysis.

Now we move to next stage—doing the regression. This analysis will inform us as to whether the relationship is statistically significant and whether we can add a line to our graph. For linear regression we are asking whether the slope of the best-fit line to these data is significantly different from zero ($H_0 : slope = 0$). To accomplish this we use the linear model function lm().

```
> mod = lm(BAC ~ drinks) # this is y ~ x
```

With this type of statistical test it's good practice to store the result in a variable, such as "mod," so that we can parse out the different pieces. We also can send that result to other functions to do things like get the p-value or draw the best-fit line on our graph. We can get most of what we need using the summary(), so let's look at what we have:

```
> summary(mod)

Call:
lm(formula = BAC ~ drinks)

Residuals:
       Min         1Q      Median         3Q        Max
-0.0032727 -0.0028636  0.0002727  0.0027273  0.0038182

Coefficients:
              Estimate Std. Error t value Pr(>|t|)
(Intercept) 0.0000000  0.0021106    0.00        1
drinks      0.0232727  0.0003402   68.42 2.32e-12 ***
---
Signif. codes:  0 '***' 0.001 '**' 0.01 '*' 0.05 '.' 0.1 ' ' 1

Residual standard error: 0.00309 on 8 degrees of freedom
Multiple R-squared:  0.9983,        Adjusted R-squared:  0.9981
F-statistic:  4681 on 1 and 8 DF,  p-value: 2.318e-12
```

From our analysis we're interested in getting the following:

1. the equation (or model) that describes the relationship $(y = f(x))$;
2. the significance of the statistical test; and
3. how well the points fit the model (the R^2 value).

From the summary above we can get all of this information. The output, when formatted like this, provides us a section under the "Coefficients," which displays the estimates for the slope, labeled "drinks," and the intercept. We also get the standard errors for these estimated values. The last column gives us the p-value for a hypothesis test for the intercept and slope. The H_0 for both of these statistics is that they are zero. We see that the estimate for the intercept is very close to 0.0 and that the resulting p-value is very close to 1.0 (to several decimal places). This strongly suggests that the intercept is not statistically different from zero. That makes sense! If a male does not have a drink his blood alcohol content should be zero.

Additionally, the output suggests that the slope is different from zero (p = 2.318 * 10^{-12}). This also is the **p-value** for the overall test for regression. When we report the results from a regression we generally need the F statistic, the degrees of freedom, the p-value, the adjusted R^2 value, and the equation, all provided in this summary.

The larger the F statistic the more confidence we have that there is a non-zero slope of the y-variable on the x-variable. The R^2 value, also known as the "coefficient of determination" (also written as r^2) is a value that ranges 0 $\leq$ The $R^2 \leq 1$. It gives us the proportion of the variance in our y variable that is explained by our x variable. In this particular case our The R^2 value is quite high (almost 1.0), which suggests a very strong fit.

If the slope (or overall regression test) is significant then we can add the best-fit regression line. The easiest approach is to use the **abline()** function, which draws a line that goes all the way from the axis on the left to the axis on the right (the two "ordinates") (see the left panel in Figure 9.5). Unfortunately, we shouldn't do this, because we need to be very careful about predicting y-values beyond the range of our x-values.

To avoid this we can, instead of using **abline()**, use the following code to draw our line, assuming we have our data in two arrays of equal length called **x** and **y**:

```
> lines(x,fitted(lm(y~x)))
```

Undoubtedly, this single line of code seems relatively simple to draw a best-fit line on some data. It has a function (**lm()**) that's in a function (**fitted()**) that's in yet another function (**lines**). After our initial shock that this is what it takes to draw a simple line, let's tear it apart, working from inside out.

The **fitted()** function takes the output from the **lm()** function and determines the predicted y values for the provided x values. Finally, the **lines()** function draws the line between all the points (it's straight because all the fitted points lie on the same function returned by **lm()**). The end product is the graph on the right of Figure 9.5.

```
> par(mfrow = c(1,2))
> plot(drinks,BAC, pch= 16, cex.lab = 1.5,
```

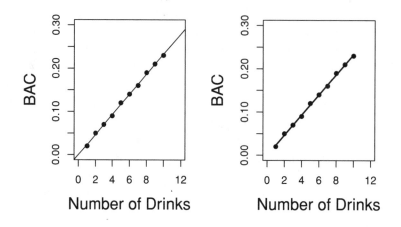

Figure 9.5: The data for blood alcohol content (BAC) graphed as a function of the number of drinks consumed by male subjects. In the left panel the line extends beyond the range of the x-variable. In the right panel the line is graphed correctly, extending only as far as the range of drinks consumed. I've asked the function to double the thickness of the line with the argument `lwd` = 2.

```
+       xlim = c(0, 12), ylim = c(0,0.3),
+       xlab = "Number of Drinks")
> abline(mod)
> plot(drinks,BAC, pch= 16, cex.lab = 1.5,
+       xlim = c(0, 12), ylim = c(0,0.3),
+       xlab = "Number of Drinks")
> lines(drinks,fitted(mod), lwd = 2)
> par(mfrow = c(1,1))
```

ANOTHER EXAMPLE OF REGRESSION ANALYSIS

I was recently reading a paper from the journal *Science* that reported how warning calls of black-capped chickadees (*Poecile atricapillus*) are influenced by the size of the predator. Chickadees are small song birds that make a call that sounds, not surprisingly, like "chick-a-dee." The researchers found that chickadees tag on extra "dee" sounds to their calls depending on the extent of a threat–the more threatened they are, the more "dee" sounds they make. Larger predators actually are less of a threat to chickadees than smaller predators so the average number of "dee" sounds *decreases* with the size of potential predators. Given this we should expect to see a graph of the number of calls going down as size of the predator goes up.

From the paper I estimated the values from the graph using the simple paint program on my computer to reproduce the graph from the paper. Here are the data that I was able to get:

```
> aveNumD = c(3.96, 4.10, 3.58, 3.05, 3.17,
+             3.23, 2.77, 2.29, 1.76, 2.21,
+             2.47, 2.79, 2.56, 2.26, 2.07,
+             1.35)
> PredLength = c(16.16, 18.87, 35.37, 31.67,
+                50.15, 51.13, 27.73, 38.33,
+                24.29, 51.13, 55.07, 53.60,
+                60.49, 59.51, 66.40, 55.32)
```

We can graph these data and verify this decreasing relationship (see Figure 9.6).

```
> plot(PredLength,aveNumD,pch = 16,xlim = c(0,70),
+      ylim = c(0,4.5),
+      xlab = "Predator Body Length (cm)",
+      ylab = "Dee Notes Per Call",
+      xaxs = "i", yaxs = "i",cex.lab = 1.5)
> fit = lm(aveNumD ~ PredLength)
> lines(PredLength,fitted(fit),lwd = 2)
> # Add the line from description of Figure2B.
> abline(a = 4.4, b = -0.4, lwd = 2,
+        lty = 2) # line from pub
```

As described in the figure caption, there are two lines fit on the graph. The solid line represents the best fit for the data. The dashed line is the equation for the relationship presented in the paper. It's important to think skeptically. I didn't know the equation was wrong until I plotted it with the data (see Figure 2B in Templeton et al., 2005).

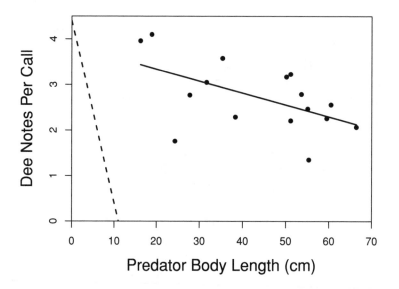

Figure 9.6: The number of "dee" call notes from chickadees in the presence of different potential predators, represented by their average lengths. The solid line is the best-fit line through these data ($y = -0.026x + 3.855$; $F = 5.865$; $df = 1, 14$; $p = 0.0296$). The dashed line is the equation presented in the paper ($y = -0.4x + 4.4$, $p < 0.0001$, $R^2 = 0.361$).

9.3 PROBLEMS

(For questions 1–4). Identify whether these examples of experiments would best be investigated using correlation or regression analysis. Also consider whether the relationship might be linear or not and whether the relationship may be positive or negative. For such problems it's usually quite helpful to actually hand-draw a graph of the expected relationship.

1. The length of metacarpals and metatarsals in whales.
2. The top speed of a car and the power of the engine, rated in "horsepower."
3. A person's life expectancy and the average number of cigarettes smoked per day.

4. Mammalian heart rate versus body mass.

(For questions 5–6). The following data are femur and humerus lengths (mm) for five fossils of the extinct proto-bird *Archaeopterix*. If they are from the same species then these points should exhibit a positive relationship.

Individual	Femur (mm)	Humerus (mm)
1	38	41
2	56	63
3	59	70
4	64	72
5	74	84

5. Do these data follow a linear relationship?
6. Graph these data and, if you decide that a line is appropriate, add the best-fit line. If not, do not add the line.

(For questions 7–9). The following are the number of breeding pairs of red-tailed hawks in a state from 2004–2009:

Year	# Pairs
2004	955
2005	995
2006	1029
2007	1072
2008	1102
2009	1130

7. Create the most appropriate visualization of this change over time.
8. If appropriate add the best-fit line to the data if, and only if, the relationship is significant.
9. If appropriate what is the rate of change in the number of breeding pairs for this species in this state?

10. Loblolly pine trees, like all organisms, grow in height over time during at least part of their life. There are data for tree growth in a built-in dataframe called "Loblolly" that contains data for the heights of trees over several years. The heights are in feet. Using data for one of the seed types of your choice test whether the height (`Loblolly$height`) is increasing over time (`Loblolly$age`). Note that you need to *subset* one of the seed types to perform such a test.

CHAPTER 10

HYPOTHESIS TESTS: OBSERVED AND EXPECTED VALUES

Sometimes we have data where we have counted the numbers of things. Such data are generally called "categorical data." A classic example we often run into in biology laboratories is to test whether phenotypes adhere to a 3:1 ratio, or a 9:3:3:1 ratio. We'll discuss the chi-square test (χ^2 test) and, when the dataset has few observations, the Fisher exact test.

10.1 THE χ^2 TEST

Imagine you have 50 yellow pea plants and 20 green pea plants. You're wondering if they adhere to a 3:1 ratio. Here's how you'd do it. We first enter our data. The easiest way is to enter our observed count data and then enter the expected *probabilities*.

```
> obs = c(50,20) # observed counts
> expP = c(0.75,0.25) # expected probabilities
```

Once we have our data entered we do our test with the following command. You might notice that I haven't done a normality test. This test does not assume the data are normally distributed. It is, therefore, a non-parametric test.

```
> chisq.test(obs, p = expP)

        Chi-squared test for given probabilities
```

```
data:  obs
X-squared = 0.4762, df = 1, p-value = 0.4902
```

The chisq.test() function returns the chi-square statistic ($\chi^2 = 0.476$), the degrees of freedom (df = 1), and the p-value (p = 0.49). The null hypothesis is that the observed values adhere to the 3:1 ratio (expP). In this example it appears that the data do follow a 3:1 ratio because p > 0.05.

We can understand this test much better by looking at the equation used to calculate the χ^2 statistic:

$$\chi^2 = \sum_{i=1}^{n} \frac{(obs_i - exp_i)^2}{exp_i} \tag{10.1}$$

The obs_i in equation 10.1 represents the count for the ith category. The exp_i represents the expected count for the ith category. In the 3:1 ratio example we have just two categories (yellow and green). If the obs_i and exp_i are the same for all categories then the χ^2 value will be zero, which would be consistent with the H_0 that the observed data adhere to the expected. As the observed and expected values increasingly differ more and more, the χ^2 statistic will get bigger. Once the χ^2 statistic reaches and/or exceeds a critical value, we will find that $p \leq \alpha$ and we will reject our H_0 that the expected and observed counts do not differ.

For the above example, where we had 50 and 20 individuals that should fall into a 3:1 ratio, we can calculate χ^2 by hand using equation 10.1 as follows:

$$\chi^2 = \frac{(50 - 52.5)^2}{52.5} + \frac{(20 - 17.5)^2}{17.5} = 0.4762$$

I determined the expected values by multiplying 0.75 and 0.25 by the total number of individuals in the sample (70).

There are a few caveats with this test. First, you need to be careful about proportions versus counts (or frequencies). The above equation uses counts while the function in R is best used with proportions. If you know it's a 3:1 ratio, for instance, then the expected frequencies should have 75% and 25% of the values, respectively. If you have 1371 observations you'd have to calculate the number for each category ($1371 \cdot 0.75$ and $1371 \cdot 0.25$). But, no matter how many total observations you have, the proportions should still be 3:1, or 0.75 and 0.25.

A second concern involves limitations to the chi-square statistical test. There should be no more than 20% of the expected frequencies less than 5 and none should be fewer than one. If this is a problem you should use the Fisher exact test (see section 10.2 on page 159).

DATA IN A CONTINGENCY TABLE

You may encounter observed and expected data that occur in a matrix, such as an n by m "contingency table." The chi-square test can easily work with such data. You need to enter your data into a matrix and send that matrix to the `chisq.test()` function. Let's try this with the following data. We might, for instance, have a sample with males and females and two hair colors (brown and blond). Assuming we have 10 brown-haired males, 6 blond-haired males, 8 brown-haired females, and 12 blond-haired females, we might be interested in whether individuals are distributed as expected across gender and hair color. This is actually a really tricky test to understand. Here are our data:

	Brown	Blond	Total
Males	10	6	16
Females	8	12	20
Total	18	18	36

I've included the totals for the rows and columns in this matrix. Those are used to calculate the expected values for the observed values. Notice that there are 16 males and 18 brown-haired people in our population of 36 individuals. Given this we can calculate the expected number of individuals which should be brown-haired males. To get the expected value we multiply the row total by the column total and divide by the grand total:

```
> 16*18/36
```

```
[1] 8
```

I got that there should be eight such individuals but we saw 10, so we saw too many! Let's ask R to do the work for us. We first enter the data into a matrix and verify they're correctly entered:

```
> mat = matrix(c(10,6,8,12), byrow = TRUE, nrow = 2)
> mat

     [,1] [,2]
[1,]   10    6
[2,]    8   12
```

Now we can do our statistical test by simply sending the matrix to the `chisq.test()` function:

```
> chisq.test(mat)

        Pearson's Chi-squared test with Yates' continuity
        correction
```

data: mat
X-squared = 1.0125, df = 1, p-value = 0.3143

So, even though there are some differences in our data they do seem to follow what we would expect, given how many people fall into the two genders and the two different hair colors. What if the data are significant? Let's look at another dataset, one you might have heard of.

Lots of kids get warts. Two-thirds of the warts go away on their own within two years. Kids think warts are gross and sometimes are willing to take extraordinary means to get rid of them. The most common method is to freeze them with liquid nitrogen (cryotherapy). Another technique is to use "tape occlusion." A 2002 study (Focht et al., 2002) tested the efficacy of placing duct tape over warts for two months. Fifty-one patients completed the study and ranged in age from 3 to 22 years of age. Each was randomly assigned to either the duct tape (DT) treatment or the cryotherapy (Cryo) treatment. Note that there is no control group. Here are the results:

Trmt	Resolved	Not
DT	22	4
Cryo	15	10

The authors of the study concluded that duct tape significantly reduced the occurrence of the warts. Let's test this ourselves. We first need to enter the data:

```
> M = matrix(c(22,4,15,10),byrow = TRUE, nrow = 2)
```

Let's visualize these data. The best way is probably to show the numbers of patients in each group using a barplot (see Figure 10.1).

```
> barplot(M,beside = TRUE, ylim = c(0,30),
+   xlab = "Treatment", ylab = "Number of Patients",
+   names = c("Resolved","Not Resolved"),
+   legend = c("DT","Cryo"), cex.lab = 1.5)
> abline(h=0)
```

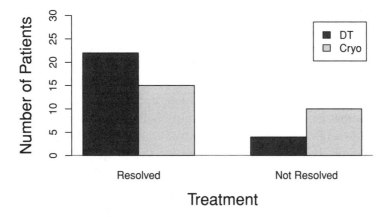

Figure 10.1: Barplot of the test for whether duct tape and cryotherapy reduced the prevalence of warts in patients.

Next we need to simply send this matrix M to the `chisq.test()` function:

```
> chisq.test(M)
```

```
        Pearson's Chi-squared test with Yates' continuity
        correction
```

```
data:  M
X-squared = 2.7401, df = 1, p-value = 0.09786
```

This is interesting. We get a p-value of 0.1. Why did the authors get a p-value of less than 0.05? The reason they got their answer is that they chose not to use a common continuity test called "Yates' correction." It turns out that R, by default, uses this correction due to the degrees of freedom in this contingency table being equal to one (df = 1). Some argue whether the correction should be used. If we tell R notto do this correction then what happens? Here you go:

```
> chisq.test(M,correct = F) # not using the Yates correction
```

```
        Pearson's Chi-squared test

data:  M
X-squared = 3.8776, df = 1, p-value = 0.04893
```

Interesting! The result is statistically significant ($p \leq 0.05$). So, do people who use duct tape lose their warts significantly more than expected, according to these data? I would say not (while others, apparently, would say so). At least we now know why the authors thought their result was significant. We have one last piece of this puzzle to evaluate, though. Was the number of kids who had warts resolved using duct tape *fewer* than expected? Here's a way to look.

First, we need to save the results from the analysis in a variable so we can pull it apart. We'll use the analysis used by the authors to see what they did.

```
> cs = chisq.test(M,correct = F)
```

Next we can see what information there is in this variable cs:

```
> names(cs)

[1] "statistic" "parameter" "p.value"   "method"
[5] "data.name" "observed"  "expected"  "residuals"
[9] "stdres"
```

We're most interested in the observed and expected values, so let's look at those:

```
> cs$obs

     [,1] [,2]
[1,]   22    4
[2,]   15   10


> cs$exp # you can use just enough letters to be unique

          [,1]      [,2]
[1,] 18.86275 7.137255
[2,] 18.13725 6.862745
```

The upper-left cell is what interests us (resolved using duct tape). We saw 22 people in this group and expected 18.86 to have their warts resolved with duct tape. That is consistent with the finding of the paper. We also notice that there were four people who used duct tape and it didn't resolve the wart. We expected 7.14. So, fewer warts than expected were not resolved. Again, this is consistent with the conclusion of the paper. It is now left to us to decide whether this is or is not a statistically significant result. Is the result from the experiment unexpected, or is it consistent with what would be expected by chance? It's up to you to determine!

10.2 THE FISHER EXACT TEST

The previously discussed χ^2 test with the continuity correction is an approximation for this test (Zar, 2009). We can run this on a matrix much like we did using the chi-square test on the duct tape experiment.

```
> M = matrix(c(22,4,15,10),byrow = TRUE, nrow = 2)
> fisher.test(M)

        Fisher's Exact Test for Count Data

data:  M
p-value = 0.06438
alternative hypothesis: true odds ratio is not equal to 1
95 percent confidence interval:
  0.8333154 18.6370764
sample estimates:
odds ratio
   3.57229
```

This simple test on these data suggest what we had discovered from the chi-square test with the Yates' correction; that the results are not statistically significant. Therefore, we conclude that using duct tape did not have an effect on the resolution of warts, compared to the cryotherapy treatment (OR = 3.57, p = 0.064).

10.3 PROBLEMS

(For questions 1–2). Corn kernels were counted in an ear of corn. There were 295 purple kernels and 86 yellow kernels.

1. Are these counts consistent with a 3:1 ratio of purple to yellow?
2. Provide a graph of observed and expected counts.

(For questions 3–6). Gregor Mendel reported (1866, see Hartl and Fairbanks, 2007) finding pea plants in one dihybrid cross experiment with the following ratio:

Phenotype	Number
Round/Yellow	315
Wrinkled/Yellow	101
Round/Green	108
Wrinkled/Green	32

3. Are these consistent with a `9:3:3:1` ratio?
4. Provide a barplot of just the data.
5. Extract from R the expected values had the plants occurred in exactly a 9:3:3:1 ratio (round the values).
6. Provide the observed and expected values in a single barplot.

(For questions 7–10). The following data represent the observed body mass index values for a male population with 708 individuals. Also included are the expected proportions for this particular population.

BMI	< 18	≥ 18 but < 25	≥ 25 but < 30	≥ 30
Observed	57	330	279	42
Expected Percentages	10	40	40	10

7. Are the number of people in the BMI categories consistent with the expected percentages for this population?
8. Provide a barplot of these data.
9. Extract from R the expected values had the plants occurred in exactly a 9:3:3:1 ratio (round the values).
10. Provide a barplot of the observed and expected values in a barplot.

CHAPTER 11

A FEW MORE ADVANCED PROCEDURES

11.1 WRITING YOUR OWN FUNCTION

So far you have used a large number of functions in R. If you look at the index under "R functions" you'll see a long list of these. Sometimes, however, you might want to write your own function. Functions are bundled code that you can write and, after running the code once, you'll be able to use it over and over.

To get started let's write a function that simply returns the arithmetic mean of an array. This isn't really necessary since R provides us with such a function (**mean()**). It's a pretty simple task which will be good for demonstration. The arithmetic mean ($\bar{x}$) for a set of data can be written as follows:

$$\bar{x} = \frac{1}{n} \sum_{i=1}^{n} x_i \tag{11.1}$$

Note that functions in programming languages like R are really the right-hand side of the equations. They first evaluate the right-hand side and then send the result to the left-hand side. We might store that result in a variable or simply code the right-hand side so that the output is printed to the console. Here's the difference using the built-in function for the mean:

```
> my.dat = c(6,3,4,5,3,1)
> answer = mean(my.dat) # store the return solution in a var.
> mean(my.dat) # no storage so print the answer to console

[1] 3.666667
```

To write a function we need to name the function and decide what the function will take, or accept, as arguments. In this example of calculating the mean, we will want our function to take an array of numbers and then return the mean. Let's call our function my.mean(). The function can be written like this:

```
> my.mean = function (x) {
+   ans = sum(x)/length(x)
+   return (ans)
+ }
```

The format for a function includes the assignment of a function to a variable of your choice. Above I have named my mean function my.mean(). It will take a single variable that I have chosen to name x. As written the variable x could be any type of object. I'm thinking that the user, however, will only send an array of numbers (hope I'm right!). All of the code within the function is wrapped in curly braces.

The first line calculates the arithmetic mean of the x variable. That's done by summing the numbers and dividing that sum by the number of values in the data array x. The answer is stored in a variable called **ans**. Finally, the function returns whatever is stored in **ans**.

To get a function like this to be usable you have run it, just like other code in R. When you run it nothing should happen! If something is written to the console it'll be an error message. So, nothing is good. After this you need to test your function—send it some data and see if it works. Let's test whether it works with this code:

```
> my.dat = c(6,5,4,5,6,5,4,1) # create some data
> my.mean(my.dat) # send the new function the data

[1] 4.5
```

Did it work? We can test it against the built-in mean() function like this:

```
> mean(my.dat)

[1] 4.5
```

Functions can be fun to write. A really good function, however, can be challenging to write. The above function my.mean() is nice but it isn't at all robust. It's quite easy to break! Try getting the mean of "Bob," for instance. A good function will check the incoming data and, if necessary, return an error message that's at least somewhat useful (you've probably already seen several error messages provided by R!). Functions are useful to write when

1. R doesn't have a function to do what you want to do;
2. you need a particularly calculation that
   (a) you need to do many times; and
   (b) you want to control that it's absolutely done correctly;
3. you want to bundle a series of tasks into a single command to make analyses easier for you and/or your collaborators (e.g., read in an updated data file, summarize it, do a statistical test, and then produce an appropriate visualization);
4. you want to share code you've developed that solves a particular problem.

Once you exit R your own function will go away. For it to be available in your next R session you'll have to run the function. Then it will become part of the R session and will be available and ready for you to use.

Let's build a function that allows us to add a proper best-fit line to a linear regression. We introduced this idea in the section on linear regression (see section 9.2 on page 143). To do this we simply wrap our code in a "function declaration" and identify the arguments we want the user to be able to send to the function. Here's a simple version of the function:

```
> lm.line = function (x, y) {
+   lines(x,fitted(lm(y~x)))
+ }
```

I'm going to create some linear but noisy data and store the result from the linear regression analysis in a variable called mod.

```
> set.seed(100)
> x = 5:20
> y = 0.4*x + 2 + 7.5*runif(length(x))
> mod = lm(y~x)
```

I can now graph these data and compare the drawing of the line using abline() versus our own function lm.line() (see Figure 11.1).

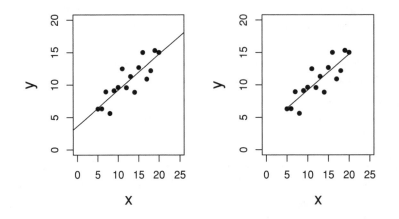

Figure 11.1: Two graphs with best-fit linear regression lines. On the left I used the standard approach with the `abline()` function that extends from axis to axis. In the right graph I used our new `lm.line()` function that extends the line only over the range of the x variable.

```
> par(mfrow = c(1,2)) # create a two-panel graphics window
> plot(x,y, xlim = c(0,25), ylim = c(0,20), pch=16,
+       cex.lab = 1.5)
> abline(mod)
> plot(x,y, xlim = c(0,25), ylim = c(0,20), pch=16,
+       cex.lab = 1.5)
> lm.line(x,y)
```

As you can see, we can now add appropriate best-fit lines to scatterplots using our `lm.line()` function.

11.2 ADDING 95% CONFIDENCE INTERVALS TO BARPLOTS

Sometimes you'll be asked to add confidence intervals to a barplot. For the method I'm introducing you need to make sure you have installed the `plotrix` package:

```
> install.packages("plotrix") # typed at the command line
```

Now you can proceed with the following set of commands to apply 95% confidence intervals to your barplots.

Assume that we have data on the heights, in meters, of three species of trees (`hts`). Our boss (or professor) has asked us to make 95% confidence intervals and add them to the barplot. This will require us to calculate the error bar sizes for each sample and then add those to the barplot. We can calculate a 95% confidence interval as follows:

$$CI_{95} = T \cdot SEM$$
$$= t_{1-\frac{\alpha}{2}, df} \cdot \frac{s}{\sqrt{n}}$$

OK, this is just intimidating. But, in R, it's not so bad! We might first recognize that this is the `t-value` (`T`) multiplied by the standard error of the mean (`SEM`). To get `T` we set $\alpha = 0.05$ and `df = n-1` and for `SEM` we just need the standard deviation `s` and the number of values in our sample (see section 4.3 on page 54).

To get `T` we use the `qt()` function, which returns the desired `T` if we send it $1 - \alpha/2$ and our degrees of freedom (df).

Let's assume that we sample the heights for $n = 10$ trees each of red oaks, white pine, and sassafras. For our trees we find that their mean heights are 12m, 15.5m and 10.2m and have standard deviations of 2.3m, 3.3m, 2.6m, respectively:

```
> N = 10
> hts = c(12,15.5,10.2) # mean heights of trees
> S = c(2.3,3.3,2.6) # standard deviations
```

To build our 95% confidence intervals we need to calculate T and SEM:

```
> T = qt(1-0.05/2,9) # for all samples if df = 10-1 = 9
> SEM = S/sqrt(10) # because S is an array, SEM is an array
```

If our sample sizes are all the same then we only have one `T` value (it's dependent on our chosen α and the number of observations in each sample, which gives us `df`). If the sample sizes are different then we would need separate `T` values for each sample (e.g., T_{RO}, T_{WP}, and T_S). The 95% CI is now just the product of these two terms ($T \cdot SEM$). We can do this for all three trees simultaneously like this:

```
> CI95 = T*SEM # our 95% CI length (one direction)
```

Once we've calculated the 95% confidence intervals, we're ready to make our graph. We can provide some names to label our bars in our barplot, make the barplot, and then add the CIs. Notice that we save the output from the barplot (`a = barplot(...)`) because this gives us our x locations for the centers of the bars in the barplot. We'll use these to place the error bars in the middle of the bars. The `abline()` call after the barplot puts a line along the x-axis. The final step, which relies on the function `plotCI` in the `plotrix` package, adds the CIs to our barplot (the argument `add = T`).

The `pch` argument defines which point character to draw where the error bar meets the top of each bar. I like the error bars to just be lines without marking the center with a point character, so I set `pch = NA` to omit the points.

```
> library(plotrix) # you need to have installed this.
> names = c("Red Oak", "White Pine", "Sassafras")
> a = barplot(hts, ylim = c(0,20),names.arg = names,
+             ylab = "Height (m)", cex.lab = 1.5,
+             xlab = "Tree Species")
> abline(h = 0)
> plotCI(a,hts,CI95, add = T, pch = NA)
```

11.3 ADDING LETTERS TO BARPLOTS

Barplots typically allow us to compare visually the means of different samples. We're usually interested in testing whether the samples are statistically different using, for instance, the Tukey Honest Significant Differences test (`TukeyHSD`). In this section we'll see how we can make these differences clear graphically.

Figure 8.1 on page 119 shows a comparison of means among three samples when the data have been represented as boxplots. Visually it seems that the samples might all be statistically different (among `low`, `med`, and `high`). As we saw back in section 8.1 on page 120 the *post hoc* Tukey HSD test provides us with a pairwise comparison among all the treatment levels after completing

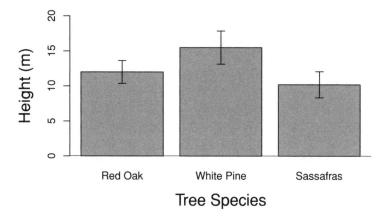

Figure 11.2: The mean heights for three tree species. Error bars represent ±95% confidence interval. Note that you can't correctly determine statistical differences among samples based on overlapping error bars. The comparison needs to be done more formally using a *post hoc* test.

an analysis of variance. Below I've reproduced the data of fish masses from chapter 8:

```
> Low = c(52.3, 48.0, 39.3, 50.8, 53.3, 45.1)
> Med = c(50.4, 53.8, 53.4, 58.1, 56.7, 61.2)
> High = c(66.3, 59.9, 59.9, 61.3, 58.3, 59.4)
> my.fish.dat = data.frame(Low,Med,High)
> my.fish.dat = stack(my.fish.dat)
> names(my.fish.dat) = c("Mass","Trmt")
> my.fish.dat$Trmt = factor(my.fish.dat$Trmt,
+        levels = c("Low","Med","High"))
> boxplot(my.fish.dat$Mass ~ my.fish.dat$Trmt,
+        names = c("Low","Medium","High"),
+        xlab = "Feeding Rate", ylab = "Fish Mass (g)",
+        cex.lab = 1.5)
```

We found these data to be normally distributed.

```
> shapiro.test(Low)$p.value # output suppressed
```

```
> shapiro.test(Med)$p.value
> shapiro.test(High)$p.value

> # output suppressed
> my.aov = aov(my.fish.dat$Mass~my.fish.dat$Trmt)
> summary(my.aov)
```

Finally, we performed the *post hoc* Shapiro-Wilk test on these data and found the following:

```
> TukeyHSD(my.aov)[1] # the [1] provides just the table
```

```
$`my.fish.dat$Trmt`
                diff       lwr       upr        p adj
Med-Low     7.466667  1.311785 13.62155 0.0170661543
High-Low   12.716667  6.561785 18.87155 0.0002184511
High-Med    5.250000 -0.904882 11.40488 0.1008029585
```

Recall that p-values $\leq \alpha$ (0.05) suggest that samples are statistically different. The output from the Tukey test suggests that the levels `low` and `high` are statistically different and `low` and `med` are statistically different. There is no statistical difference between `med` and `high` ($p = 0.10$).

What biologists often do is provide barplots with 95% confidence intervals and, if a *post hoc* test has been done, provide letters above the samples to show statistical differences. Placing letters is a bit complicated so we might even want to do this with pencil and paper to begin. I like to just hand draw the graph and make connections between means (vertical bars) that are not statistically different. Once you have this, then you need to assign letters to each of the bars. Means that are not statistically different must share the same letter.

The usual approach to assigning letters is to proceed in alphabetical order, starting with "a," which is assigned to the lowest mean and letters assigned upward with increasing means. It is possible for a mean to have two or more letters. The goal, however, is to use as few letters as necessary. For the relatively simple example with just three bars, the assignment of letters is relatively easy (see Figure 11.2 on page 167).

Once we have created our barplot and added 95% CIs, then the letters, once determined and provided in correct order, can be added to the barplot with the following two lines:

```
contrast.labels = c("a","b","b")
text(a, M+CI95, labels =   contrast.labels, pos = 3)
```

The first line above assigns the letters to an array. They should be in the order of the means as they appear in the barplot. The second line writes them just above the top of the error bars, in the proper place. The `text()` function writes text on a graph at the x-y coordinates that you provide.

In this example we provide the `text()` function two arrays for the x and y coordinates. The x-coordinates come from the `barplot()` function, which returns the x-coordinates for the middles of the bars graphed. The y-coordinates are heights in the graph. I add to the means (`M`) the height of the individual confidence intervals (`CI95`). The `labels` are the letters we want to assign to each bar (from left to right).

The last argument for the `text()` function is the position (`pos`). The choices for this are the integers 1–4, with these representing below, left, above, and right, respectively. We want the letters on top, so have provided the `text()` function the additional argument "`pos = 3`." Sometimes in biology we don't provide the error bars and, instead, simply provide the means in a barplot and use letters to represent statistical differences among means. This is easily done by not adding the error bars and using just the mean heights of the bars and `pos = 3` in the `text()` function. Adding letters to a barplot like this is appropriate only if the statistical test is significant.

> Note: You may not perform a *post hoc* test or add letters to barplots if the overall analysis of variance is not statistically significant, indicating that the samples appear to come from the same population. Therefore, they are not statistically different and should not have different letters.

Below is the code used to make Figure 11.3, which shows both 95% error bars and letters from a *post hoc* test.

```
> library(plotrix) # you need to have installed this.
> M = tapply(my.fish.dat$Mass,my.fish.dat$Trmt,mean) #get means
> S = tapply(my.fish.dat$Mass,my.fish.dat$Trmt,sd) # get sds
> M = c(M[2],M[3],M[1]) # rearrange the order (low, med, high)
> S = c(S[2],S[3],S[1]) # rearrange the order (low, med, high)
> T = qt(1-0.05/2,5) # for all samples if df = 6 - 1 = 5
> SEM = S/sqrt(6) # because S is an array, SEM is an array
> CI95 = T * SEM
```

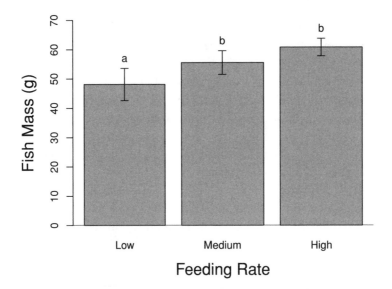

Figure 11.3: The mass of fish grown under three food treatments. The samples are significantly different ($F = 202; df = 2, 15; p < 0.001$). These are the same data shown in Figure 8.1 on page 119. Means with the same letter are not statistically different.

```
> a = barplot(M, ylim = c(0,70),
+             names.arg = c("Low", "Medium", "High"),
+             ylab = "Fish Mass (g)", cex.lab = 1.5,
+             xlab = "Feeding Rate")
> abline(h = 0)
> plotCI(a,M,CI95, add = T, pch = NA)
> contrast.labels = c("a","b","b")
> text(a, M+CI95, labels =   contrast.labels, pos = 3)
```

11.4 ADDING 95% CONFIDENCE INTERVAL LINES FOR LINEAR REGRESSION

When we conduct a regression analysis *and* get a significant result *and* the relationship between the x and y variables is linear, we can add a best-fit straight line. When we look at the analysis output we see that our slope and intercept estimates have error terms. This is because the data points do not all fall exactly on the best-fit line. We can represent this uncertainty in regression with 95% confidence interval lines on either side of our best-fit line. This can help a reader interpret the strength of the regression relationship.

Let's start with a dataset that's built into R called "BOD," for biochemical oxygen demand. You can look at these by typing the dataset's name at the console. To add the 95% confidence interval lines we need to conduct the linear regression analysis with the linear model function (lm())) and send the output from that function to the predict() function. This returns our 95% prediction values as a matrix that we send to the lines() function. So, we make the scatterplot, add the best-fit line, and then add the confidence lines (see Figure 11.4).

```
> plot(BOD$Time,BOD$demand, pch = 16, ylim = c(0,30),
+        ylab = "BOD (mg/l)", xlab = "Time")
> mod = lm(BOD$demand~BOD$Time)
> lm.line(BOD$Time,BOD$demand) # line function from earlier
> newx = BOD$Time# seq(min(BOD$Time),max(BOD$Time))
> prd<-predict(mod,#newdata=data.frame(x=newx),
+              interval = c("confidence"), level = 0.95,
+              type="response")
> lines(newx,prd[,2],lty=2, lwd = 2)
> lines(newx,prd[,3],lty=2, lwd = 2)
```

11.5 NON-LINEAR REGRESSION

This section introduces more complicated models to fit to your data. If you are thinking of fitting a curved line to your data you should convince yourself that your data satisfy the following criteria:

1. Your data clearly are not linear and can't be transformed to become linear.

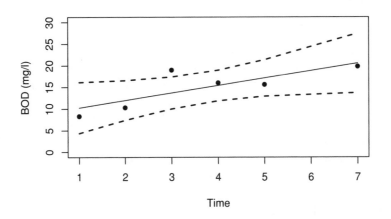

Figure 11.4: Scatterplot of the data with a best-fit linear regression line added. In addition, the 95% confidence lines for this relationship have been added.

2. The functional form that your data seem to adhere to can be represented by an equation that defines (not describes) the dynamics of the system.

Note that you are not fishing for some curvy line that looks good with these data! You need to understand how your data are related (e.g., how y depends biologically on x) and fit that functional form to your data.

This would be a good time to follow Albert Einstein's recommendation about finding an appropriate model: it should be as simple as possible, and no simpler. When we have data we often are interested in finding the best model that fits our data. That means we really want the equation that provides us our parameter estimates (remember, "statistics" are estimates of "parameters"— see section 6.1). If the equation properly fits our data, then we have a model that describes the underlying biological process. Interestingly, nature actually often is relatively simple (i.e., there are relatively few parameters to estimate and, therefore, the equations aren't too complicated).

A LINEAR EXAMPLE: DOLPHIN GROWTH

A quick example should help (see Figure 11.5). Assuming we're interested in getting the functional relationship of y on x, then we should use the function that makes the most sense biologically (represent's reasonably how the system changes over time).

Here are growth data for the mass (kg) of a stranded dolphin over a 10-week rehabilitation period.

```
[1] 87.0 86.2 89.3 89.6 88.5 89.2 90.8 90.3 92.0 92.2
```

We don't want to try to fit the highest possible polynomial to our data just so the line goes through several (or every!) point. A linear growth rate over this relatively short, 10-week period seems reasonable (see left panel in Figure 11.5)

Just for fun I used the equation I got from the 9th–order polynomial fit in the graph on the right of Figure 11.5. Isn't this better? This line goes through every data point exactly! If I use this model, however, to predict the mass of this dolphin in the 11th week, the model says it will weigh -48.1 kg. That isn't a very useful prediction! This model does not describe the growth of dolphins. Below is the code that creates Figure 11.5.

```
> mod1 = lm(y~x)
> mod2 = lm(y ~ poly(x,9))
> par(mfrow = c(1,2))
> plot(x,y, xlim = c(0,12), ylim = c(85,95),
+       ylab = "Dolphin Mass (kg)", xlab = "Time (weeks)",
+       pch = 16, cex.lab = 1.5)
> lines(x,fitted(mod1),lwd = 2)
> plot(x,y, xlim = c(0,12), ylim = c(85,95),
+       ylab = "Dolphin Mass (kg)", xlab = "Time (weeks)",
+       pch = 16, cex.lab = 1.5)
> xvals = seq(1,10,by=0.1)
> yvals = predict(mod2,list(x = xvals))
> lines(xvals,yvals,lwd = 2)
```

EXAMPLE: HEIGHTS OF PLANTS

Figure 11.6 shows the heights of plants grown in a greenhouse over the course of 16 months. The plants do not appear to be growing linearly in height. I

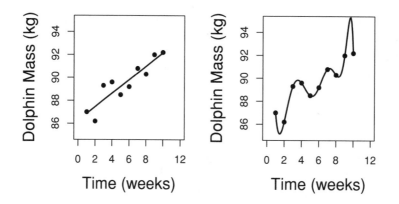

Figure 11.5: The change in average mass of a dolphin over 10 weeks. In the left panel we see the mass with a linear fit to the data. On the right I fit a 9th-order polynomial equation. It's a great fit because the line goes through every data point, but it is biologically meaningless!

have fit a straight line to the data but we can see clearly that this line does not capture how these plants are growing.

```
> ht = c(4.0, 8.5, 12.2, 13.4, 15.0, 17.8, 19.3, 19.4, 21.2,
+        21.7, 23.4, 23.8, 24.1, 24.7, 24.9, 25.5)
> time = 1:16
> mod = lm(ht ~ time)
> plot(time,ht, xlim = c(0,16), ylim = c(0,30),
+      ylab = "Plant Height (cm)", xlab = "Time (months)",
+      pch = 16, cex.lab = 1.5)
> lines(time,fitted(mod), lwd = 2)
```

Why isn't this a good model? It captures the basic increase of the data. However, we notice a clear pattern of the lack of fit. On both ends of the fit the model over-estimates the values of the data. Likewise, in the middle of the line the model underestimates the data. This tells us that this model clearly is not describing the behavior of our system.

We are interested in the coefficients that describe the growth of plants but this is going to be more complicated than simply finding the best-fit straight

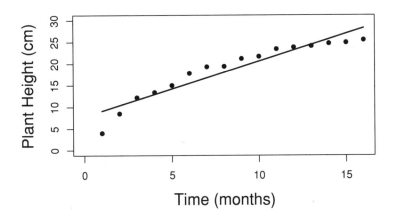

Figure 11.6: The heights of plants do not appear to increase in a linear fashion over time.

line. We need to develop a model that properly defines the growth of these plants over time. A starting point might be to think of the height of a plant at time = 0. At that time then height probably is zero as well. Therefore, our function should go through the origin.

So, how do we find the appropriate function and then get R to tell us what the equation is? To do this we have tell R what the equation is, give R some starting values that are reasonable for the coefficients, and then run the statistical test. If we've done our part correctly R will kindly return the best-fit equation and enable us to draw that function on our graph.

Finding the right equation is hard. This usually comes from researchers with experience in the system and/or from previous work on plants. Once we have an equation we then need to estimate the coefficients of the equation to help R get started searching for the best-fit coefficients. Since the data curve, let's try a simple quadratic function (a second-order polynomial). We can use the curve() function, which we first saw back in chapter 1.

```
> plot(time,ht, xlim = c(0,16), ylim = c(0,30),
+       ylab = "Plant Height (cm)", xlab = "Time (months)",
+       pch = 16, cex.lab = 1.5)
> curve(-0.12*x^2+3.1*x+3,from = 1,to = 16,add = TRUE)
```

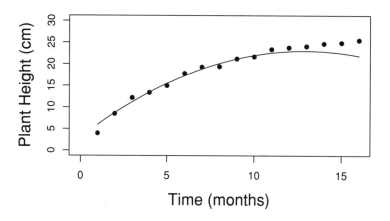

Figure 11.7: A polynomial equation fit to the data (see equation 11.2). The line does a pretty good job of going through the data points, but can you see a problem lurking in the fit?

I tried a few combinations of numbers with the curve function to get a relationship that resembled the data. I don't need to be really close but the right functional form is important. Here's a function with some coefficients that do a fairly good job.

$$y = -0.12 \cdot x^2 + 3.1 \cdot x + 3 \qquad (11.2)$$

This line, shown in Figure 11.7, isn't a bad fit to the data but it has a **serious** problem. Can you see the problem? The curve fits the data pretty well, except near the end. Is that a problem? You bet! We're hoping to use an equation that describes the underlying mechanism that governs the growth of these plants. The equation that I have used fits the data well but completely misses *how* the plants grow. This equation suggests that the trees will get smaller and eventually have a negative height. Do trees, in general, start to get shorter over time? Although we have done a great job fitting a beautiful line through these data, the underlying, quadratic model is terribly wrong!

We need, instead, to find an equation that starts at the origin (seeds at time zero have approximately zero height) and grows asymptotically. We can use a relatively simple, two-parameter model that goes through the origin and

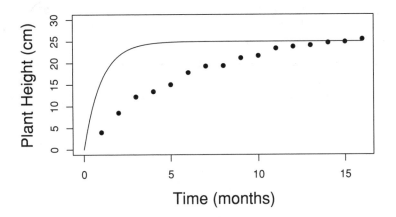

Figure 11.8: An asymptotic function that is a poor fit to the data but represents a relationship that seems to have the right idea.

reaches an asymptote:

$$y = a \cdot e^{-bx}$$

For this function we need to tell R some approximate values for the coefficients a and b. If we again look at the data we might see that the asymptote seems to be around 25. For our function the asymptote is a. We can also see that if we make $x = 0$ then the intercept is zero. So perhaps $a = 25$. I don't know what b is so let's just start it at $b = 1$. We can add this curve on the graph of the data (see Figure 11.8 on page 177).

```
> plot(time,ht, xlim = c(0,16), ylim = c(0,30),
+      ylab = "Plant Height (cm)", xlab = "Time (months)",
+      pch = 16, cex.lab = 1.5)
> curve(25 * (1 - exp(-(1*x))),from=0,to=16,add = TRUE)
```

We can see that our function gets to the asymptote too quickly (Figure 11.8). But it does have a form that includes an asymptote and is heading in the right direction. Let's ask R to approximate the best-fit values for our model:

```
> fit=nls(ht~a*(1-exp(-b*time)),start=list(a=25,b=1))
```

The nls() function conducts the non-linear, least-squares analysis. It needs the mathematical function we're trying to fit and our best-guess estimates for the coefficients. We send these as a "list."

If our calling of the function doesn't return an error then our model results are stored in the variable fit. We can now send fit to the function summary() and see the parameter estimates and the estimates of variability for these:

```
> summary(fit)

Formula: ht ~ a * (1 - exp(-b * time))

Parameters:
  Estimate Std. Error t value Pr(>|t|)
a 26.828212   0.439173   61.09  < 2e-16 ***
b  0.176628   0.007396   23.88 9.59e-13 ***
---
Signif. codes:  0 '***' 0.001 '**' 0.01 '*' 0.05 '.' 0.1 ' ' 1

Residual standard error: 0.5266 on 14 degrees of freedom

Number of iterations to convergence: 6
Achieved convergence tolerance: 7.27e-07
```

From this we can see estimates of the coefficients and error estimates for each coefficient. The equation is, therefore:

$$height = 26.8 \cdot \left(1 - e^{-0.177 \cdot time}\right)$$

In the output we also see that R has provided the p-values for each parameter estimate. In this example we see that the estimate of each parameter is "highly significant" ($p < 0.001$). This is important to us because we need to verify that the coefficients we've asked R to fit with our equation are actually important. If we "overfitted" our model (had too many parameters) we would find non-significant coefficients. This would suggest to us that we should consider a simpler model.

This equation is exciting because it gives us some insight as to how our plant changes in height over time. We also can compare our results with results by other researchers. We can test whether the parameters change given

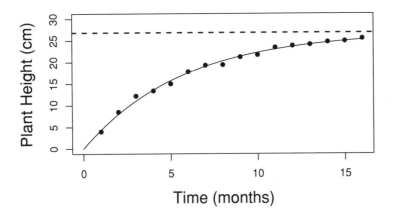

Figure 11.9: A best-fit asymptotic function to our data for the change in the height of a plant over time. The horizontal line above the curve is our asymptote.

different environmental conditions. And we can ask what the instantaneous rate of change is, using calculus! All sorts of great things become possible! Here's the final graph with the best-fit line (Figure 11.9).

```
> plot(time,ht, xlim = c(0,16), ylim = c(0,30), pch = 16,
+    ylab = "Plant Height (cm)", xlab = "Time (months)",
+       cex.lab = 1.5)
> av = seq(0,16,0.1)
> bv = predict(fit,list(time = av))
> lines(av,bv)
> asymptote = sum(coef(fit)[1]) # asymptote is the
>      # sum of these two coefficients
> abline(h = asymptote, lwd = 2, lty = 2)
```

We let R calculate the value of the asymptote by using the **coef()** function, which extracts just the estimated values of the coefficients from the model. We store that result in the variable called, not surprisingly, **asymptote**. We then send that calculated version to the **abline()** function and have it add a horizontal reference line (hence the **h = asymptote** argument) to our graph.

I've asked for a red line, although you're probably seeing this in black and white.

GET AND USE THE DERIVATIVE

In biology we often have situations where something is changing over time or as a function of some other variable. We graph our points and wish to fit a function to the data to learn more about what is going on, or to make some prediction. One such problem commonly encountered is with the Michaelis-Menton relationship. Many introductory laboratories include a module on looking at enzyme kinetic reactions as a function of substrate concentration.

Here are the steps, or the algorithm, for finding the rate at any point along a Michaelis-Menton function derived from data:

1. Get the data into two arrays of equal length.
2. Create a scatterplot of the data.
3. Find the best-fit function for these data using the Michaelis-Menton equation v = Vmax*[S]/(Km + [S]).
4. Add the best-fit line to the graph.
5. Get the derivative of the Michaelis-Menton function.
6. Find the slope of the tangent at a give value of [S].
7. Add the tangent line to the best-fit curve for the data.
8. Report the slope of the tangent line (the rate) and this given value of [S].

Here are some sample data for the substrate concentration (S) and the velocity of the reaction (v).

```
> S = c(0.2,0.5,1,2,5) # x-axis data ([S])
> v = c(0.576, 0.762, 0.846, 0.860, 0.911) # velocity

> plot(S,v, xlab = "[S]", ylab = "Velocity", xlim = c(0,max(S)),
+       ylim = c(0,1),cex.lab = 1.5, pch=16)
```

We can plot these data with a simple call using the `plot()` function (see Figure 11.10). To determine the best-fit line we use the non-linear least-squares function (`nls()`), which returns the coefficients for the model. We must, however, tell R what model to use. In this case we'll specify the Michaelis-Menton function $\left(v = \frac{Vmax \cdot fit \cdot S}{Km \cdot fit + S}\right)$. R, like all statistics programs, requires us to provide educated guesses of the coefficients we're fitting. Note

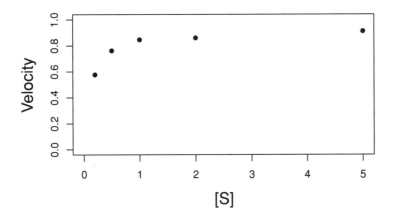

Figure 11.10: A scatterplot of the data for velocity and substrate concentration.

that our unknowns are V_{max}.fit and K_m.fit. R will use our starting guesses to begin its search for the best parameter estimates.

```
> mod = nls(v~(Vmax.fit*S)/(Km.fit + S),
+           start=list(Vmax.fit = 1,Km.fit = .1))
```

If it works without returning an error, our resulting model will be stored in the variable **mod**. We can now add the best-fit line that R has given us to the data in our graph. Sounds tough, but it's not in R. We can look at the model by typing "**mod**" at the command prompt and hitting <enter>. If, however, we send **mod** to the **summary()** function, we get a better and more easily understood summary of the model output:

```
> summary(mod)

Formula: v ~ (Vmax.fit * S)/(Km.fit + S)

Parameters:
          Estimate Std. Error t value Pr(>|t|)
Vmax.fit 0.930469   0.011879   78.33 4.59e-06 ***
```

```
Km.fit    0.118509    0.009646    12.29   0.00116 **
---
Signif. codes:  0 '***' 0.001 '**' 0.01 '*' 0.05 '.' 0.1 ' ' 1

Residual standard error: 0.01536 on 3 degrees of freedom

Number of iterations to convergence: 4
Achieved convergence tolerance: 4.272e-07
```

We can see that the two parameters have estimates and variabilities (standard errors). Let's use the model output to draw the best-fit line onto the graph with the data (see Figure 11.11). We use the **predict()** function to give us the model's y-values and draw that line over the range of x-values.

```
> plot(S,v, xlab = "[S]", ylab = "Velocity", xlim = c(0,max(S)),
+        ylim = c(0,1),cex.lab = 1.5, pch=16)
> x = seq(0,max(S),by = 0.01) # an array seq. of x values
> y = predict(mod,list(S = x)) # the predicted y values
> lines(x,y, lwd = 2) # draw the line
> asymptote = coef(mod)[1] # value of the asymptote
> abline(h = asymptote)
```

GET THE SLOPE OF THE TANGENT LINE

We can get the slope at any point on our function by taking the derivative of our function. As we have seen above, the coefficients are saved in the variable "mod" so we can get these parameter estimates individually. Let's get them from the mod and just call them a and b for simplicity. To do this we can use the function **coef()**.

```
> Vmax = coef(mod)[1] # this is Vmax
> Km = coef(mod)[2] # this is Km
```

Now we need the derivative to the Michaelis-Menton equation. We tell R what the equation is and then get the derivative with the function D(). I'm going to choose a value of [S] where I want the tangent line to intercept my function. I'll call that particular value my.S. I don't want to use the original variable "S" because that will over-write the original data.

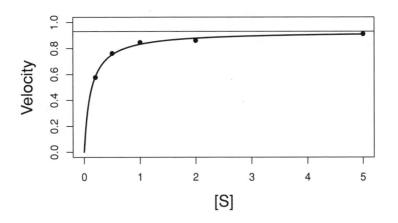

Figure 11.11: A scatterplot of the data for velocity and substrate concentration with the best-fit line. The equation for the line, returned by the `nls()` function, is $v = \frac{0.930 \cdot S}{0.118 + S}$.

```
> my.exp = expression(Vmax * my.S/(Km+my.S))
> my.deriv = D(my.exp,"my.S")
> my.deriv # look a the derivative returned by D()

Vmax/(Km + my.S) - Vmax * my.S/(Km + my.S)^2
```

The next step we need to take is to find the tangent line at [S] = `my.S`. To do this we need to find the slope at a particular value of [S], such as at `my.S` = 1. We can also add a vertical line at this point to help convince ourselves we're finding the tangent at this point along our function (see Figure 11.12).

```
> plot(S,v, xlab = "[S]", ylab = "Velocity", xlim = c(0,max(S)),
+        ylim = c(0,1),cex.lab = 1.5, pch=16)
> my.S = 1 # choose where on [S] to get the rate (tangent line)
> der.slope = eval(my.deriv) # get slow using deriv. at my.S
> der.y = eval(my.exp) # Ht of the best-fit function at my.S
> der.int = der.y - der.slope*my.S # The intercept of tangent
> lines(x,y, lwd = 2)
```

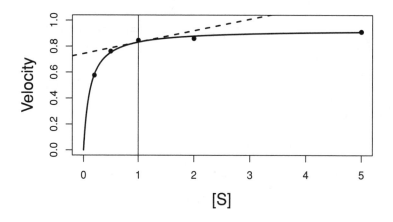

Figure 11.12: A scatterplot of the data for velocity and substrate concentration with the best-fit line. The equation for the line, returned by the `nls()` function, is $v = \frac{0.930 \cdot S}{0.118 + S}$ with a tangent line added at [S] = 1.

```
> asymptote = Vmax
> abline(der.int,der.slope,lwd = 2,lty = 2) # draw tangent
> abline(v=my.S) # place a vert line at tangent
```

Finally, let's have R report the slope of the tangent at this point for our chosen [S].

```
> cat("The velocity of the reaction at",
+     my.S," is ",der.slope,"\n")
```

The velocity of the reaction at 1   is   0.08814004

I suggest you try to draw that tangent line in another place and get the rate at that point.

11.6 AN INTRODUCTION TO MATHEMATICAL MODELING

In addition to being able to conduct statistical tests we sometimes need to create and evaluate mathematical models. We might, for instance, be interested in using the statistical fit from data and then use those parameter estimates to build a model, such as exponential growth for a population. We'll look at this model and then at a more complicated, standard epidemiology model.

EXPONENTIAL GROWTH

Exponential growth is a relatively simple differential equation model:

$$\frac{dN}{dt} = rN \tag{11.3}$$

This model defines the instantaneous rate of change of the population (dN/dt) as a function of a growth rate parameter (r), times the population size N. This equation defines the rate of change of the population. But what is the population size at any given time? To determine this we need to "solve" this differential equation, which means we need to integrate it. If we integrate this model we find the solution:

$$N_t = N_0 e^{rt} \tag{11.4}$$

In practice we are unable to solve most differential equations analytically and find their solutions, like above. Instead, we have to rely on programs like R to solve differential equations *numerically*. Let's solve this relatively simple differential equation model numerically to see how this is done in R. We're going to use some code that is admittedly challenging to understand. We'd love to take a bunch of time to figure this out but we're going to just implement the approach here. Through this example, and the next, you'll be able to solve a variety of other problems with minor changes to this code.

So, instead of using the analytical solution (equation 11.4) we'll ask R to approximate the solution from equation 11.3. We can do this as follows:

```
> library(deSolve)  # library for solving diff. equations
> num_yrs = 10
> r = 0.2 # the growth rate parameter
> N0 = 100 # starting population
> xstart = c(N=N0) # create a "list" of constants -
>     # used in the solve function
> parms = r # here's our one parameter
> mod = function(t,x,parms) {
+    N = x[1]
+    with(as.list(parms) , {
```

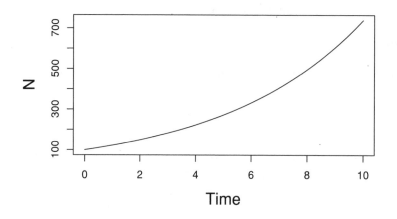

Figure 11.13: The graphical solution to exponential growth (equation 11.3).

```
+       dN = r*N
+       res=c(dN)
+       list(res)
+    })}
```

Once we have this entered and run this code in R we will have the model (mod) stored in the dataframe "output."

```
> time=seq(0,num_yrs, by = 0.1)  # set number of time steps
> # RUN THE MODEL in the next line!
> output = as.data.frame(lsoda(xstart,time,mod,parms))
> plot(output$time,output$N, xlab = "Time", ylab = "N",
+       type = "l", cex.lab = 1.5)
```

The last step is to simply plot the data from the model. R provides this in the variable output from which we access the x- and y-axis data. We then ask R to plot() the data and add the argument type = "l" (that's the letter "el"), which produces the smooth curve for the population growth (see Figure 11.13).

THE SIR MODEL

In our second model we're going to see the outcome from a standard epidemiology model, called an "SIR" model for **S**usceptible, **I**nfectious, and **R**ecovered individuals. Here's the model, called a "system of differential equations" by mathematicians and a "coupled differential equation model" by biologists. The bottom line is that these equations depend on each other because they share state variables (S and I).

$$\frac{dS}{dt} = -\beta SI$$
$$\frac{dI}{dt} = \beta SI - \nu I \qquad (11.5)$$
$$\frac{dR}{dt} = \nu I$$

Each equation governs the dynamics of each of these groups of individuals. The model is technically "closed" since no individuals enter or leave the model. Instead, individuals can simply move from being susceptible to being infectious to being recovered. Susceptible people leave that class at rate β if they come into contact with an infectious individual. Thus, the term for $\frac{dS}{dt}$ is negative $(-\beta SI)$. Those individuals that leave the S state enter the infectious class. Infectious individuals recover at rate ν.

The system of equations (equation 11.5) is clearly intimidating to biologists! But it is easy to solve using R. What we're after as biologists is how the prevalence of a disease might change over time. In particular, if you have a bent toward being interested in the health sciences, we'd like to see this disease disappear. If we measure infection and recovery rates we might find that we get the dynamics seen in Figure 11.14.

```
> Num_Days = 20  # number of days to run simulation
> B = 0.006      # transmission rate (0.006)
> v = 0.3        # recovery rate (0.3)
> So = 499          # initial susceptible pop (299)
> Io = 1            # initial infectious pop (1)
> Ro = 0            # initial recovered pop (0)
> xstart = c(S=So,I = Io, R = Ro)
> parms = c(B, v)
```

The code below defines the model as a **function**. The first part sets the initial values for S, I, and R, which are sent to the function as the x arguments. The with() function contains the actual mathematical model (equation 11.5).

```
> mod = function(t,x,parms) {
+   S = x[1] # init num of susceptibles
+   I = x[2] # init num of infectious
+   R = x[3] # init num of recovered
+   with(as.list(parms) , {
+     dS = -B*S*I        # dS/dt
+     dI = B*S*I - v*I   # dI/dt
+     dR = v*I           # dR/dt
+     res=c(dS,dI,dR)
+     list(res)
+   })}


> times=seq(0,Num_Days,length=200)  # set up time steps
> # RUN THE MODEL!
> output = as.data.frame(lsoda(xstart,times,mod,parms))
> plot(output$time, output$S, type="n",ylab="Abundance",
+       xlab="Time", main="ODE Model", cex.lab = 1.5,
+       ylim = c(0,So*1.6))
> leg.txt = c("S","I","R")
> legend("topright",leg.txt,lwd=2,lty = 1:3)
> lines(output$time,output$S,lty = 1,lwd=3)
> lines(output$time,output$I,lty = 2,lwd=3)
> lines(output$time,output$R,lty = 3,lwd=3)
```

We can now run the model, much like we did for the exponential growth model. The actual running of the model is done with the single line that starts with "output = ." It seems pretty simple but what R is doing behind the scenes is really amazing (and complicated)! It is approximating the solution to the system of differential equations and storing the result in the dataframe output. The remaining code creates the plot and draws the S, I, and R populations using the lines() function.

11.7 PROBLEMS

1. Write a function called my.stats() that takes an array of numbers and gathers the mean, standard deviation, and standard error of the mean

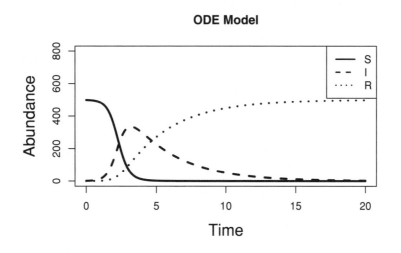

Figure 11.14: The dynamics of an SIR model with $\beta = 0.006$ and $\nu = 0.3$. We note that the infectious class (I) (dashed line) increases rapidly and then turns downward and approaches zero. This is referred to as an "epidemic curve." The susceptible class (solid line) eventually decreases to near zero. Finally, over the epidemic, we see nearly all individuals end up in the recovered (R) class (dotted line).

    into a dataframe and returns it. Test your function with data and verify that your function returns the correct values.

2. The following data are blood clotting times in minutes for a drug administered to laboratory rats.

Drug	Blood clotting times (min)
A	8.8, 8.4, 7.9, 8.7, 9.1, 9.6
B	10.6, 9.8, 10.1, 8.4, 9.6, 10.2
C	11.6, 11.4, 9.1, 10.7, 14.9, 12.9

Create a barplot of these data, including 95% confidence intervals on each sample.

3. Using the data in the question above (problem 2), perform an analysis of variance test on the different drugs and provide a graph of the results using a barplot with the error bars representing the standard error of the means for each sample. Perform a Tukey HSD *post hoc* analysis on these

data, if the ANOVA is statistically significant, and provide letters on top of the error bars identifying statistical differences among the samples.

4. Using the built-in Loblolly pine dataset (see problem 9.3 on page 151), graph the change in the height of trees of seed type 301. The trees seem to exhibit linear growth over time. If the relationship is significant add the best-fit line, using linear regression, and finish by adding the 95% confidence interval lines to the graph.

(For questions 5–6). Below are data for one car's gas mileage in miles per gallon (mpg) determined instantaneously at different speeds in miles per hour.

mpg	24.1	28.1	33.1	33	31
speed	20	30	40	50	60

Assume that the function that defines these data is a simple, second-order quadratic function (e.g., $y = a \cdot x^2 + b \cdot x + c$).

5. Create a publication-quality graph of these data with the best-fit quadratic function added to the data.
6. What is the optimal speed for the car to travel in order to maximize gas mileage? To complete this exercise you'll need to rely on calculus.

(For questions 7–8). Consider the following "logistic growth function"

$$\frac{dN}{dt} = rN(1 - \frac{N}{K})$$ (11.6)

which has the following analytical solution:

$$N_t = \frac{K \cdot N_0 \cdot e^{r \cdot t}}{K + N_0 \cdot (e^{r \cdot t} - 1)}$$ (11.7)

where the initial population is $N_0 = 50$, $r = 0.82$, and $K = 1000$.

7. Use R's solver (deSolve) to find and graph the solution to the differential equation (11.6) over the period from $t = 0 - 10$.
8. Use the curve() function to plot the solution (equation 11.7) from $t = 0 - 10$.

(For questions 9–10). Enzyme kinetics can be described with the following Michaelis-Menton equation:

$$v = \frac{V_{max} \cdot [S]}{K_m + [S]}$$

where v is the velocity of the reaction, V_{max} is the maximum velocity, $[S]$ is the concentration of the substrate, and K_m is the Michaelis-Menton constant.

9. The built-in dataset called `Puromycin` contains data on the reaction rate versus substrate concentration for cells treated and untreated with Puromycin. Create side-by-side scatterplots for the "treated" and "untreated" data. [Hint: you'll need to `subset()` the data.]

10. Add to each scatterplot the best-fit relationships.

CHAPTER 12

AN INTRODUCTION TO COMPUTER PROGRAMMING

In addition to the statistical power of R you will find that R also is a full-featured, object-oriented programming language. In general, we use programming to get the computer to perform tasks that need to be done many times. Our problems can be quite complex and so we also need to control the problem-solving route to include conditional statements, such as "if this is true then do task 1, otherwise do task 2." This is done a lot in biology for a variety of problems. This introduction is meant to give you a taste of what is possible. Knowing what's possible can allow you to think of new problems that you might consider solving.

12.1 WHAT IS A "COMPUTER PROGRAM"?

Computers need to be told exactly what you want them to do. And they are really good at doing what we tell them to do (like when it's time to show a user the "blue screen of death"). And they're particularly good at doing things repetitively.

For our programming needs computers run sets of instructions. These instructions are written by people like you and me that use a "high-level" language that makes sense to us. Other programs, written by other humans, then take the instructions and convert them into "low-level" instructions that computers understand. My goal here is to introduce you to the high-level language of R so you can solve problems in biology.

There are many high-level languages that can be used to create computer programs. You may have heard of some of these (e.g., "Matlab," "Maple," "Mathematica," "Basic," "Visual Basic" [used within Excel], "Pascal," "Python," and "Perl"). Different languages have their own strengths and weaknesses. Strangely, there is no *"best"* language. However, for modeling complex biological systems most any problem can be solved using either C, a relatively low-level language, or R, or a combination of these two languages.

C is great for relatively complicated problems that need to run fast. R, on the other hand, is easy to use but runs programs much slower than C. The advantage of R, however, is that there is a very large set of easy-to-use, built-in tools for creating graphs, running statistical tests, and analyzing mathematical models. Therefore, although R programs run slowly the "development time" for writing programs in R is very short compared to writing programs in C. So, it really helps if you know what you're trying to do before you invest a lot of time on your research project. Keep in mind, however, that if you can program in one language it becomes relatively easy to migrate to a new language.

Since this document is just an introduction we won't be tackling a complex problem in biology. You will have to continue to explore and read about your problem and programming. I hope, however, that you can get started and see why you might want to do this and how programming can be used to understand and predict patterns in biology.

Computers are great at doing lots of calculations that we'd find pretty boring after a while. There are three basic constructs we use in building programs:

1. `if` statements (conditional tests)
2. `for` loops (control flow)
3. `while` loops (control flow)

The `if` statement is used to test whether a condition is true. If the condition evaluates to true, then the following statement is executed. If the conditional test is false, then the true expression is ignored. You can provide an expression to execute if the statement is false with an **else** expression.

We use `for` and `while` loops when we want a particular expression executed more than one time. The `for` loop should be used if the number of times something needs to be done is known (we usually know this). If the number of times a statement is to be executed is unknown, then it might be best to use a `while` loop. The `while` loop is done until some criterion is satisfied.

AN EXAMPLE: THE CENTRAL LIMIT THEOREM

Let's build a short program that tests the central limit theorem (CLT), which states that the distribution of sample means from any distribution is approximately normal. To accomplish this let's sample the standard normal distribution ($\bar{x} = 0$ and $\sigma = 1$) and a uniform distribution ($x \in [0, 1]$) 5000 times each. Since we know how many times we want to do the sampling it is most appropriate to use a **for** loop.

To test the CLT we will perform a simulation which requires us to use a random number generator. Let's begin by setting the random seed so we'll get the same results:

```
> set.seed(20)
```

We can then declare our sample size (**num.values**) and the number of times we take samples from each population (**num.samples**) like this:

```
> num.values = 100
> num.samples = 5000 # the number of samples to draw
```

For each of the 5000 samples we need to calculate and store the mean of the 100 values in each sample. We, therefore, need two structures that will each hold a list of 5000 mean values. For this we'll use arrays. Arrays can be declared as follows

```
> means.norm = numeric(num.samples) # will hold 5000 values
> means.unif = numeric(num.samples)
```

At last we are ready to perform our simulation and collect our data (stored in the **means.*** variables):

```
> for (i in 1:num.samples) {
+    means.norm[i] = mean(rnorm(num.values))
+    means.unif[i] = mean(runif(num.values))
+ }
```

The above code does all the work. It tells R to run the two lines between the curly braces ({ }) 5000 times (from 1 to **num.samples**). The first line within the **for** loop works like this:

1. send **num.values** (100) to the **rnorm()** function;

2. the `rnorm()` function returns 100 random numbers drawn from the standard normal distribution;

3. those values from the standard normal distribution are then sent to the `mean()` function;

4. the `mean()` function calculates the arithmetic mean of those numbers and assigns that result to the ith element in the `means.norm[]` array.

Once the for loop is done the variables `means.norm` and `means.unif` should hold 5000 values each. We now can look at the distributions of these means using the `hist()` function (see section 5.2 on histograms). I've also plotted the distributions of single samples from the standard normal and uniform distributions for comparison (see Figure 12.1).

```
> par(mfrow = c(2,2))
> hist(rnorm(num.samples),xlab = "X",main = "Normal Sample",
+       ylim = c(0,1000))
> hist(runif(num.samples),xlab = "X",main = "Uniform Sample",
+       ylim = c(0,1000))
> hist(means.norm, xlab = "X",main = "Means from Normal Dist",
+       ylim = c(0,1000))
> hist(means.unif,xlab = "X",main = "Means from Uniform Dist",
+       ylim = c(0,1000))
> par(mfrow = c(1,1))
```

Now we can view the output by displaying four graphs in a single graphics window (Figure 12.1). Interestingly, the histograms in the lower part of the figure appear similarly bell-shaped. Remember that the histogram in the lower right is from 5000 means of samples drawn from a non-normal (uniform) distribution. We can test them for normality using the Shapiro-Wilk test (see section 4.4):

```
> shapiro.test(means.norm) # from normal pop

        Shapiro-Wilk normality test

data:  means.norm
W = 0.9997, p-value = 0.7198


> shapiro.test(means.unif) # from unif pop
```

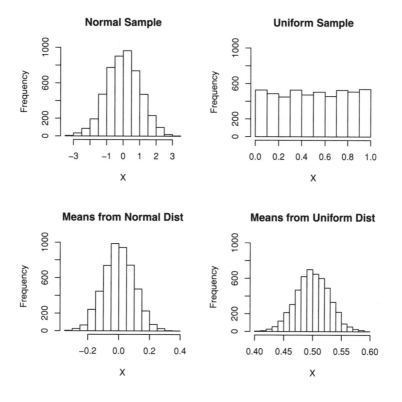

Figure 12.1: A visualization of the central limit theorem. The upper panels show the distribution of two samples of 5000 deviates, one drawn from a normal distribution (top left) and one drawn from a uniform distribution (top right). The bottom panels demonstrate that the means of many samples drawn from a normal (lower-left panel) and uniform (lower-right panel) distribution result in relatively normal distributions, in agreement with the central limit theorem (CLT).

```
        Shapiro-Wilk normality test

data:  means.unif
W = 0.9996, p-value = 0.39
```

What do we conclude about these distributions? Does this result agree with or contradict the central limit theorem?

What we did above was to write a computer program. As we discussed above, we sometimes want to have a computer test an idea that might require many calculations. This is a great example that has allowed us to verify empirically (that means actually doing it) a theorem. Our next step is to explore further how programming can be used to answer some questions in the biological sciences.

12.2 INTRODUCING ALGORITHMS

Let's approach our introduction to algorithms with an example of simulating "genetic drift" operating on two alleles found at a single locus. Genetic drift is a mechanism that brings about evolution in a population, or the change in gene frequencies. With this mechanism, unlike natural selection, there is no selection at all. The process is governed by chance so that one allele might become more or less frequent over time simply by chance.

For this exercise let's assume that individuals are "haploid" at a single locus with just two alleles (we'll use the numbers 0 or 1 for our two alleles). Using numbers like this for our alleles will make our lives easier. Imagine that our population has six individuals (e.g., 0, 1, 1, 0, 0, 0). To know the number of 1 alleles all we need to do is add up the numbers (e.g., > sum(0,1,1,0,0,0), which means there must be two of the type 1 alleles in the population). Therefore, the proportion of allele 1 in this example is 0.333.

We'll assume that reproduction is really simplistic. The population size will remain the same size over time and that reproduction is completed by randomly choosing N individuals from the current population. The chosen individuals simply make a baby that's identical to itself which gets placed into the next generation.

What *should* happen over time? We might, just by chance, select more individuals with the 0 allele than the 1 allele over time and end up with a population of only the 0 alleles. Just as likely we might end up with the 1 allele becoming the only allele in the population. Because this is actually complicated, we need to develop an "algorithm," which is a recipe that guides us in writing our program to simulate this process. Here are the steps that I follow to complete this simulation (there are many other ways you could do this):

1. Declare the necessary variables (e.g., N);
2. Create an empty plot for our results;
3. Run the simulation many times (sounds like a "for" loop). In each simulation we:

(a) Create a population;

(b) Store the proportion of 1s in a variable (e.g., P) for each time step;

(c) Perform the reproduction routine, randomly choosing individuals that will contribute to the next generation and do this many times;

(d) Add a line for the P data to graph;

(e) If not done, repeat for next replicate.

We now need to implement these steps, assuming that the algorithm will work. What I do is write the algorithm in my script file and then write the code directly below each task. I can then test each task individually to make sure it's doing what I think it should. If the tasks are more complicated I might write a function to do the work. The code below creates Figure 12.2. Note that the first line calls the **set.seed()** function (with 10 as the argument) so that if you run this code you should get exactly the same graph as I did. (Remember that you need to omit the ">" and "+" symbols.)

```
> set.seed(10) # gives repeatable result
> # Step 1: Define variables
> n.time.steps = 100 # how long to run simulation
> pop.size = 50 # how many individuals there are
> n.reps = 20 # how many populations to simulate
> # Step 2: Create an empty plot for simulation
> plot(0,ylim = c(0,1),xlim = c(1,n.time.steps), type = "n",
+       ylab = "Proportion of Allele 1 in Population",
+       xlab = "Time Step", cex.lab = 1.5)
> abline(h=0.5,lty = 2, lwd = 3)
> # Step 3: Perform the reproduction routine many times;
> for (i in 1:n.reps) {
+    # Step 3a: Create population of 0 and 1 alleles
+    pop = c(rep(0,pop.size/2),rep(1,pop.size/2))
+    # Step 3b: Store proportion of 1s in variable
+    P = sum(pop)/pop.size
+    # Step 3c: Run the simulation for this replicate
+         for (j in 2:n.time.steps) {
+                 pop = sample(pop,pop.size, replace = T)
+                 P[j] = sum(pop)/pop.size
+         }
+    # Step 3d: Add line to graph for this replicate
+    lines(P)
+ }
```

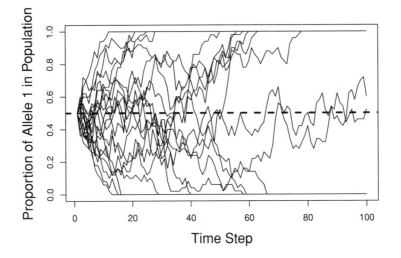

Figure 12.2: The proportion of allele 1 in 20 replicate populations over time. Each line represents a separate population. Eighteen of the 20 replicates result in either the loss or fixation of allele 1. Two simulations continue with both alleles persisting in these populations.

12.3 COMBINING PROGRAMMING AND COMPUTER OUTPUT

Sometimes we would like to have code that solves a problem and provides the answer to whomever is looking at the screen (ourselves or others). We can use commands that will provide the answer with some regular English mixed in. This can be great to help out lab mates, for instance. Here's how we might solve a small problem with a contingent result and print the appropriate result to the console.

Imagine that you want to provide your lab group with a routine that checks to see if the data they collected are normally distributed. Here's a way to do this using the `cat()` function.

```
> # x = c() # should enter data here
> set.seed(100) # so your numbers match mine
> x = rnorm(100) # using these data for demonstration
```

```
> a = shapiro.test(x)
> {
+ if(a$p.value > 0.05)
+   cat("The data are normally distributed: p = ",a$p.value)
+ else
+   cat("The data are not normally distributed: p = ",a$p.value)
+ }
```

```
The data are normally distributed: p =  0.5349945
```

The above code performs the Shapiro-Wilk test, stores the result, and then queries whether the p-value is or isn't greater than 0.05. Whichever is true, the program will spit out the answer in a nice written format.

The tools we have used above represent really important components for creating programs in R. With these tools, and perhaps a few more that you'll pick up as you need them, you'll be able to answer some really interesting questions in biology. Test yourself by trying the problems in the next section. Have fun! The key to success in computer programming is play. If you're interested in doing more programming, check out the many available sources on using R. One good, complete guide is the book by Matloff (2011).

12.4 PROBLEMS

1. Create a program that uses a **for** loop to count the integers 1:100.

(For questions 2–3). The Fibonacci series begins with the numbers 0, 1, and continues as the sum of the two previous numbers.

2. Write a program that generates the first 10 numbers of the series.
3. Create side-by-side graphs (in a single graphics window) of this series. On the left present the series on a linear scale; on the right present it on a semi-log (ln) scale (y is logged, x is linear).

(For questions 4–7). Assume that zebra mussels (an invasive, freshwater gastropod that is expanding its range in North America) are capable of growing geometrically according to the following difference equation.

$$N_{t+1} = N_t \cdot \lambda$$

A new population of 100 individuals has just arrived in a lake. In a neighboring lake, the population grew at a rate of 45% per year. Therefore, a population of 100 will be 145 in the next year. (What is the value of λ? Hint: solve for λ.)

4. Write a program that models the change in the population in this lake over the next 9-year period. Note that $N_1 = 100$ and the $N_2 = 145$. You'll want to model this up to N_{10}, which is a total of 10 population sizes.

5. Create a graph of how you expect this population to change over this time period.

6. A control method could be implemented that would reduce the growth rate by 10% annually. Show a graph of both the uncontrolled population and the controlled population for comparison, as though you were asked to show a town board what the difference in the zebra mussel population might look like over time.

7. Change the model so that it relies on the following generalized expression to the geometric growth equation:

$$N_t = N_0 \cdot \lambda^t$$

8. We've spent time in this text thinking about p-values (you might want to revisit the definition for the p-value on page 94). Write a program that tests whether this works for the standard t-test. Below is a suggested algorithm to solve this problem.

   (a) Create two samples from normal distributions. One should have $\bar{x}_2 = 0$, sd = 1; the other $\bar{x}_1 = 1$, sd = 1.

   (b) Perform a t-test on your two samples (see page 111).

   (c) Store your t- and p-values in two variables (e.g., my.t.value and my.p.value).

   (d) Write a program that iteratively, using a **for** loop, generates two samples, performs a t-test, records the t-value, and counts how many of the t-values are as extreme as or greater than your t-value from your original test. Do this 1000 times. Note that you do not want to write each number down! Have R record the number of times the new t-value is greater than your original t-value.

   (e) Print to the console your original p-value and the proportion of times you got a t-value as great as or greater than the original t-value. Your proportion and the p-value from the first test should be similar.

9. The following equation is called the "logistic map."

$$N_{t+1} = \lambda \cdot N_t \cdot (1 - N_t)$$

When λ lies in the range $3.57 < \lambda < 4.0$ the series can produce chaotic dynamics. Create a graph of what happens when $\lambda = 3.75$ over 100 time steps. Begin with $N_1 = 0.5$. [Note that $0 \leq N \leq 1$]

10. Ramanujan, a famous Indian mathematician who died at the age of 32, proposed this estimate for π.

$$\pi = 1 / \left(\frac{2\sqrt{2}}{9801} \sum_{k=0}^{\infty} \frac{(4k)!(1103 + 26390k)}{(k!)^4 396^{4k}} \right)$$

(a) Determine the estimates for π for $k = 0 \to 5$.

(b) Graph your estimates of π as a function of k.

CHAPTER 13

FINAL THOUGHTS

I hope you've learned a lot! As we discussed, this science stuff is challenging. Perhaps you learned that R is challenging, but manageable. Hopefully you also have learned that you can solve problems using R. You've learned how to install this program, and to install a front end (RStudio). You've learned how to extend this program with a few of the thousands of available packages. And I hope you've learned how to use R and RStudio, and how to use these to get data from Excel and the Internet.

Many people have gone through what you have. And most biologists nowadays have encountered or used R. Many have said "I'm not doing that" (only to quietly come on board). But you didn't say that—you've done it! You've had to look the beast in the eyes and come out victorious (at least sometimes, I hope). I commend you for your work. You should keep this skill up. You can use R for simple calculations (what's the $\sqrt{5}$?). You can use it to make a quick graph of a function as well as professional graphs, solve differential equations, and solve just about anything quantitative you can think of. Put this skill on your resumé, and continue to solve problems.

Most important, I hope you've learned to be skeptical. Skepticism means you're not sure about something until you've seen the data. Good scientists are skeptical. Don't be cynical, which is the act of rejecting an idea because it exists. And I hope you've learned to be careful believing what people say. Belief plays no role in science. So take those data, view them graphically, and test hypotheses yourself because you have access to and know how to use R, the statistical and programming choice of biologists worldwide.

13.1 WHERE DO I GO FROM HERE?

I hope you're interested in learning more about how to solve problems in biological sciences. There are large numbers of biostatistics textbooks and books that introduce and extend various skills using R. Books on more advanced statistical techniques using R are coming out weekly.

Ultimately, there are many great sources of information: in books (see Introduction) and free online sources that you should be able to use to answer any question you might face. If you have worked your way through this book and given the exercises the good ole college try then you're ready to handle the solutions you'll find online. You should be able to target your search and succeed. I often do a Google search by starting off with the letter "r" and then typing in some keywords. I rarely come up empty-handed with this approach.

There are a variety of topics I have been unable to introduce. An important frontier of modern biology has been collectively referred to as "omics," which include genomics, epigenomics, proteomics, and metagenomics, to name a few. One important tool for analysis of these systems is the package "BioConductor" (see http://www.bioconductor.org/), which runs within the R environment (see Swan, D. 2011. Analyzing microarray data in BioConductor. The Bioinformatics Knowledgeblog. http://bioinformatics.knowledgeblog.org/2011/06/20/analysing-microarray-data-in-bioconductor/). Your work on developing how to solve these problems will serve you well in your career.

If you find the quantitative analysis of data and the building of models interesting and valuable you're on your way to thinking like a modern biologist who seeks to better understand the complexity of biological systems. I recommend that you consider opportunities in the quantitative sciences. There are a variety of different opportunities to pursue as an undergraduate, including getting internships and research positions in what are called "research experiences for undergraduates," or "REU's" (see `http://www.nsf.gov/crssprgm/reu/reu_search.cfm`).

One last point. To achieve success will require hard work. No one starts answering challenging questions in science and says "this is so easy!" Those who are, or have been, successful however, might make it look easy. It's not! Everyone of those successful people worked really hard. You too can do this. It takes focus and it takes time. But the payoffs will be many. You're on your way! Good luck!

ACKNOWLEDGMENTS

I would like to thank the developers and maintainers of the open-source software that was used to develop this book. I wrote this book in a form of LaTeX (an open-source typesetting environment) with embedded R code (more open-source software). The original document is in a "noweb" format (*.Rnw) that I edited from within RStudio (yet another open-source software package) and compiled using the `Sweave()` function. With a single keystroke combination in RStudio the R code is run and written to a LaTeX file which is run through MiKTeX (more open-source software) to create the final pdf. Some of the R code relies on "packages" which were written by volunteers within the vast R community.

> What an amazing species we are to provide such rich analytical
> and writing tools to strangers for free.

I would like to thank members of the SUNY Geneseo community for their support, particularly our Provost Carol Long for the sabbatical I used to complete this book. I also would like to thank Jenny Apple, Tom Reho, Bob Simon, John Haynie, Rob Feissner, Jarrod LaFountain, Hayley Martin, Patrick Asselin, and Chris Leary for their keen eyes and insightful suggestions as to how to improve this book. I also would like to thank the many anonymous reviewers of this book who took time to help improve this book. Thanks go to my editors Patrick Fitzgerald, Bridget Flannery-McCoy, Anne McCoy, Lisa Hamm, and Ellie Thorn with Columbia University Press for their patience, insight, and belief that I could complete this project. I want you to know that I really tried!

I also must thank the students, particularly Colin Kremer and Noah Dukler, who have helped me see the challenges and rewards of learning and using R and for their patience to see me through the development of my teaching methods. Without their questions and perplexed looks this book would be far more frustrating for you than it already is. Despite all this help there undoubtly remain many shortcomings in this book, to which I enjoy complete ownership.

On a more personal note, I thank my children for being willing to not see me as much as they would have otherwise seen me (although that might have been a plus for them). And, most of all, I thank Meredith Drake for her support and amazing cat-wranglin' skills. Without you guys I'd undoubtedly be roaming Earth in search of love, hope, and purpose. And you would be living an existence without a quantitative biology book dedicated to you. So, this one's for you!

SOLUTIONS TO ODD-NUMBERED PROBLEMS

CHAPTER 1

1.   a. > sqrt(17)
        [1] 4.123106
     b. > log(10,base = 8)
        [1] 1.107309
     c. > x = 3 # first, set x = 3, then use it
        > 17 + (5*x + 7)/2
        [1] 28

3. > r = 13.5/2 # get the radius
   > pi * r^2

   [1] 143.1388

5.   a. > # Small dataset so use c():
        > my.dat = c(3.2,6.7,5.5,3.1,4.2,7.3,6.0,8.8,5.8,4.6)
     b. > mean(my.dat)
        [1] 5.52
     c. > sd(my.dat)
        [1] 1.809113
     d. > sd(my.dat)^2
        [1] 3.272889
        > var(my.dat)
        [1] 3.272889

7.   a. > 230/28 # miles per gallon (mpg)
        [1] 8.214286

b. > *(230/28)*32 # mpg * number of gallons held by tank*
   [1] 262.8571

9. Below is the code without showing the graph.

   > *Vmax = 0.9*
   > *Km = 0.12*
   > *curve(Vmax * x/(Km + x),0,5)*

CHAPTER 2

1. > *grades = c(86.3, 79.9, 92.4, 85.5, 96.2, 68.9)*

3. You should enter these data so they resemble those below.

	A
1	Mass
2	2.17
3	1.53
4	2.02
5	1.76
6	1.81
7	1.55
8	2.07
9	1.75
10	2.05
11	1.96
12	

5. No answer for this question.
7. > my.coons = read.csv("raccoons.csv")
   > names(my.coons)
9. No answer for this question.

CHAPTER 3

1. > *cheetahs = c(102, 107, 109, 101, 112)*

3. > *order(cheetahs)*

   [1] 4 1 2 3 5

5. > *signif(cheetahs,2)*

   [1] 100 110 110 100 110

7. > read.csv("mussels.csv")

9. > *my.dat[order(my.dat$values),]*

```
      values  ind
1        12  Low
4        19  Low
3        22  Low
14       22  Med
7        25  Low
5        27  Low
6        31  Low
2        32  Low
9        34  Med
13       44  Med
10       45  Med
8        54  Med
17       59 High
19       64 High
11       69  Med
20       73 High
21       77 High
16       78 High
18       82 High
12       83  Med
15       87 High
```

It's a bit challenging to see the pattern but yes, the lows seem to sort early and the highs late.

11. > *subset(my.dat,values < 30)$ind*

```
[1] Low Low Low Low Low Med
Levels: High Low Med
```

This lists, after the [1], that those samples with values less than 30 had treatment levels that were low, except for one medium sample.

CHAPTER 4

1. > *flag = c(11.1, 9.9, 8.7, 8.6, 7.0, 4.3)*
   > *cat("The mean = ",mean(flag),"\n")*

```
The mean =  8.266667
```

```
> cat("The standard deviation = ",sd(flag),"\n")

The standard deviation =  2.380476

> cat("The median = ",median(flag),"\n")

The median =  8.65

> SEM = sd(flag)/sqrt(length(flag))
> cat("The SEM = ",SEM,"\n")

The SEM =  0.9718253

> cat("The IQR = ",IQR(flag),"\n")

The IQR =  2.2
```

3.
```
> shapiro.test(flag)

        Shapiro-Wilk normality test

data:  flag
W = 0.9526, p-value = 0.7609
```

The test suggests the data are normally distributed. However, we need to be concerned because the sample size is small. It is possible that prior knowledge of the distribution of these data allows us to assume the data are or are not normally distributed.

5.
```
> my.ht = 68 # Hartvigsen's height in inches
> my.dat = rnorm(1000, mean = 68, sd = 5)
> a = boxplot(my.dat) # not showing histogram
> a$out # these are the outliers

[1] 52.89593 84.52076 51.39609 82.43667 53.59632 83.46125
```

7. Remember to set your working directory to where you saved your data.

```
> library(UsingR)
> pop = read.csv("worldpop.csv")
> head(names(pop)) # I see R puts an X in front of year

[1] "Country.Name" "Country.Code" "X1960"
[4] "X1961"        "X1962"        "X1963"
```

```
> my.dat = pop$X2012 # I can deal
> ln.my.dat = log(my.dat) # let's log these (natural log)
> par(mfrow = c(1,2))
> simple.eda(my.dat)
> simple.eda(log(my.dat))
> par(mfrow = c(1,1))
> shapiro.test(log(my.dat))
```

```
        Shapiro-Wilk normality test

data:  log(my.dat)
W = 0.9845, p-value = 0.009354
```

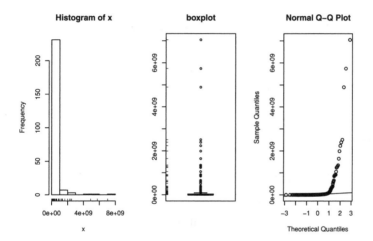

This suggests the population sizes of different countries in 2012 approaches a log-normal distribution.

9. No answer provided.

11.
```
> my.dat = runif(1000)
> IQR(my.dat)
```

```
[1] 0.4884438
```

CHAPTER 5

1.
```
> women
```

3.
```
> ?precip
> length(precip)
```

5.
```
> my.rand.dat1 = rnorm(100,mean = 10, sd = 3)
```

```
   > hist(my.rand.dat1)
7. > par(mfrow = c(1,2)) # 1 row, 2 column graphics panel
   > hist(my.rand.dat1)
   > boxplot(my.rand.dat1)
   > par(mfrow = c(1,1)) # return graphics window to one panel
9. > control = c(2,3,4,5,6,7)
   > trmt = c(5,3,4,5,6,9)
   > boxplot(control, trmt, cex.lab = 1.5,
   +         names = c("Control","Treatment"),
   +         xlab = "Treatments", ylab = "Number of Colonies",
   +         main = "Bacterial Colony Density")
```

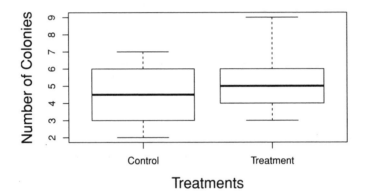

CHAPTER 6

1. If $\alpha \geq 0.06$ then $p = 0.06$ would be considered statistically significant. Notice that we need to use the "$\geq$" symbol and not "$>$".

3. This is a "type I" error. The researcher has erroneously rejected a null hypothesis. This is similar to getting a "false positive" on a drug test.

5. Here's what your graph could look like if the watering rates had an effect on plant height. The value of this exercise is getting used to imagining what you might get from an experiment before doing all the really hard work. This also leads us to consider how we would analyze these results (such as using an ANOVA test; see section 8.1 on page 117).

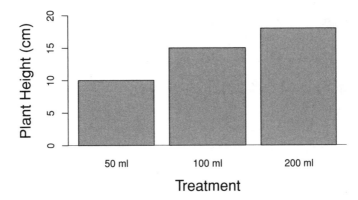

7. Inference extends only as far as the particular clone. If the researchers want to say something about cottonwood tree saplings, then they need to have several different clones. Their results do not extend beyond the clone and don't even extend beyond saplings. We are unsure of how this applies to older trees.

9. The results might extend to females. This is where, as a student, you might consider conducting the same study on female rats, or other mammal species. As the results stand they extend only as far as the male rats tested. These might have been all members of the same "line" of rats and might not apply to other rat lines or rat species. It's also questionable to consider and report on "beliefs" in science.

CHAPTER 7

1.
```
> Ah = c(1916,1563,1436,6035,3833,5031,13326,3130,6020,1889)
> hist(Ah) # graph not shown
```

This suggests these data are not normally distributed.

3. This is a two-sided test for whether 20,000 appears to come from the same population as the dataset.

```
> wilcox.test(Ah, mu = 20000)

        Wilcoxon signed rank test

data:   Ah
```

```
V = 0, p-value = 0.001953
alternative hypothesis: true location is not equal to 20000
```

This suggests that 20,000 does not come from the same population.

5.
```
> lf1 = c(27, 26, 20, 24, 28, 29)
> lf5 = c(30, 34, 28, 35, 42, 40)

> shapiro.test(lf1 - lf5)$p.value
[1] 0.6960162
Paired t-test

> t.test(lf1,lf5,paired = T)

data:  lf1 and lf5
t = -5.9656, df = 5, p-value = 0.001894
alternative hypothesis: true difference in means is
   not equal to 0
95 percent confidence interval:
 -13.116599  -5.216734
sample estimates:
  mean of the differences
-9.166667
```

These data are paired because the leaves in the first and fifth positions are on the same plants–they are not *independent*. To test for normality we look at the difference between these values. Note that the order of the data is important (the first numbers in both lists come from the same plant). From this analysis we see the data are normally distributed. We proceed with the t-test, which tells us that there is a significant difference between the leaf areas of the first and fifth leaf positions ($t = 5.96$, df = 5, p = 0.002).

7.
```
> U1 = c(81.0, 80.1, 86.1, 78.9, 86.8, 84.6, 79.3,
+          84.0, 95.4, 70.3, 86.8, 78.1)
> U2 = c(94.4, 76.7, 70.0, 88.8, 73.7, 86.3, 85.7,
+          74.0, 79.5, 75.9, 68.1, 75.9)
> shapiro.test(U1)$p.value # normally distributed

[1] 0.700756

> shapiro.test(U2)$p.value # normally distributed

[1] 0.5034876
```

```
> t.test(U1,U2, var.equal = T)

Two Sample t-test
data:
U1 and U2
t = 1.207, df = 22, p-value = 0.2403
alternative hypothesis: true difference in means is
    not equal to 0
95 percent confidence interval:
-2.537768
9.604434
sample estimates:
mean of x mean of y
82.61667
79.08333
```

The result from the t-test suggests the heights of the players are not statistically different (t = 1.207, df = 22, p = 0.24).

9.
```
> College = c(1330, 1320, 1350, 1370, 1390, 1470, 1340,
+              1470, 1450, 1360)
> University = c(1190, 1160, 1140, 1390, 1360, 1320, 1150,
+                1240, 1380, 1180)
> boxplot(College,University,
+         names = c("College","University"))
> shapiro.test(College)

        Shapiro-Wilk normality test

data:  College
W = 0.8626, p-value = 0.08197

> shapiro.test(University)

        Shapiro-Wilk normality test

data:  University
W = 0.8565, p-value = 0.06943

> var.test(College, University)

F test to compare two variances
```

```
data:
College and University
F = 0.3256, num df = 9, denom df = 9, p-value = 0.11
alternative hypothesis: true ratio of variances
    is not equal to 1
95 percent confidence interval:
0.08087545 1.31088010
sample estimates:
ratio of variances
0.3256041


> t.test(College,University,var.equal = T)


Two Sample t-test
data:
College and University
t = 3.6345, df = 18, p-value = 0.001896
alternative hypothesis: true difference in means
    is not equal to 0
95 percent confidence interval:
56.54082 211.45918
sample estimates:
mean of x mean of y
1385
1251
```

I've entered the data, visualized them (not shown), tested them for normality (they both are consistent with a normal distribution), and performed the variance test (variances are not different). I'm ready to do a t-test and conclude that students in the colleges scored statistically higher on the SAT exam than students from the universities (t = 3.63, df = 18, p-value = 0.002).

CHAPTER 8

1. Assuming you have entered the data from the table into an Excel spreadsheet and read those data into a variable called `milk.dat`, we can test for normality using each column of data separately (no need to use the `subset()` function). To do the analysis, however, the data will need to get stacked (see question #3 below).

```
> milk.dat = read.csv("milkdat.csv")
> shapiro.test(milk.dat$Farm.1)$p.value # same for others
```

```
[1] 0.9734414
```

3.
```
> milk.dat2 = stack(milk.dat)
> names(milk.dat2) = c("Bac.count","Farm")
> # Two ways to get the Farm 1 bacterial count data
> # milk.dat2[milk.dat2$Farm == "Farm 1",]$Bac.count
> # subset(milk.dat2, Farm == "Farm 1")$Bac.count
> milk.aov = aov(milk.dat2$Bac.count ~ milk.dat2$Farm)
> summary(milk.aov)
```

```
                Df Sum Sq Mean Sq F value  Pr(>F)
milk.dat2$Farm   4  803.0  200.75   9.008 0.00012 ***
Residuals       25  557.2   22.29
---
Signif. codes:  0 '***' 0.001 '**' 0.01 '*' 0.05 '.' 0.1 ' ' 1
```

Based on this analysis, the samples do not come from the same population. Therefore, they are statistically different ($F = 9.0$; df $= 4, 25$; $p < 0.001$).

5.
```
> M = tapply(milk.dat2$Bac.count, milk.dat2$Farm,mean)
> barplot(M, ylim = c(0,35), cex.lab = 1.5,
+         xlab = "Farm",ylab = "Bacterial Count (cfu/ml)",
+         names = c("Farm 1","Farm 2","Farm 3",
+                   "Farm 4","Farm 5"))
> abline(h=0)
```

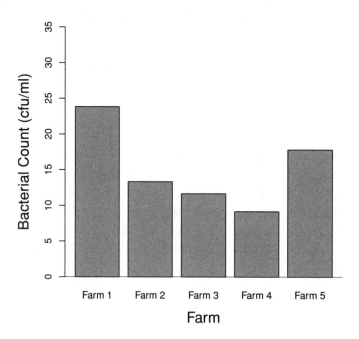

7. > my.sticks = read.csv("sticklebacks.csv")
   > a = subset(my.sticks, Temp == "C" & pH == "H")$Length
   > shapiro.test(a)$p.value # repeat for other samples

   [1] 0.492481

9. > par(mfrow = c(1,2))
   > M = tapply(my.sticks$Length,my.sticks$Temp,mean)
   > barplot(M,xlab = "Temperature",ylab = "Length (cm)",
   +   ylim = c(0,6), names = c("Cold","Warm"),cex.lab = 1.5)
   > abline(h=0)
   > M = tapply(my.sticks$Length,my.sticks$pH,mean)
   > barplot(M,xlab = "pH",ylab = "Length (cm)",
   +   ylim = c(0,6), names = c("High","Low"),cex.lab = 1.5)
   > abline(h=0)
   > par(mfrow = c(1,1))

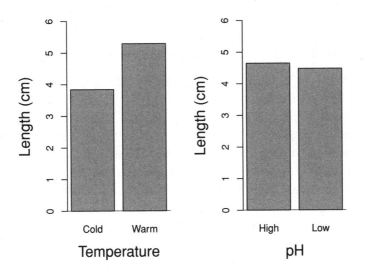

CHAPTER 9

1. Positive correlation.
3. Negative regression.
5.
```
> femur = c(38,56,59,64,74)
> humerus = c(41,63,70,72,84)
> plot(femur,humerus)
> cor.test(femur,humerus)

        Pearson's product-moment correlation

data:  femur and humerus
t = 15.9405, df = 3, p-value = 0.0005368
alternative hypothesis: true correlation is not equal to 0
95 percent confidence interval:
 0.910380 0.999633
sample estimates:
      cor
0.9941486
```

Yes, and the correlation is strongly positive ($t = 15.9$, df $= 3$, $p < 0.001$).
7. Here's the code. I've suppressed the graph itself because you can get the same graph with this code.

```
> year = 2004:2009
> pairs = c(955, 995, 1029, 1072, 1102, 1130)
> plot(year,pairs,pch = 16, cex = 1.5, cex.lab = 1.5,
+   ylim = c(0,1200), # sometimes your reader wants this
+   xlab = "Year",ylab = "Number Breeding Pairs")
```

9. To get this we need the slope of the best-fit line. We might note that the relationship doesn't quite look linear and, probably should be reaching an asymptote (curving up to some maximum). Therefore, be really careful with such a question. Generally, your professor's probably just asking for the slope of the best-fit line, which is easy. If the relationship is curved, however, then the rate (slope) changes over time. You'd get the rate at a particular time by finding the equation of the best-fit line and determining the derivative and solving for the year for which you seek the rate. Given all this, here's how to get the slope to the best-fit line, by using the lm() function and extracting the second coefficient from the returned equation (the first coefficient is the intercept).

```
> my.slope = lm(pairs~year)$coef[2]
> cat("The rate is ",my.slope,"pairs per year.\n")
```

The rate is 35.4 pairs per year.

We've answered the question as to the rate, but the equation to this line is worth discussing. Note that if you get the two coefficients the equation is:

$$pairs = 35.4 \cdot year - 69983$$

Therefore, at year zero we "*predict*" there would be −70000 breeding pairs or so. First, this suggests we shouldn't make predictions beyond the scope of our x-variable. Second, we notice that the variable *year* is on an "interval scale" but not on a "ratio scale" (see section 4.1 on page 47).

CHAPTER 10

1. ```
> obs = c(295,86) # observed counts
> expP = c(0.75,0.25) # expected probabilities
> chisq.test(obs, p = expP)
```

    Chi-squared test for given probabilities
```

```
data:  obs
X-squared = 1.1977, df = 1, p-value = 0.2738
```

Yes, the observed kernel coloration is consistent with a 3:1 ratio ($\chi^2 = 1.20$, df $= 1$, p $= 0.27$).

3.
```
> obs = c(315,101,108,32)
> expP = c(9/16,3/16,3/16,1/16)
> chisq.test(obs,p = expP)
```

```
        Chi-squared test for given probabilities
```

```
data:  obs
X-squared = 0.47, df = 3, p-value = 0.9254
```

5.
```
> a = chisq.test(obs,p= expP) # store return values
> names(a) # find which variable contains expected values
```

```
[1] "statistic" "parameter" "p.value"   "method"
[5] "data.name" "observed"  "expected"  "residuals"
[9] "stdres"
```

```
> round(a$exp,0) # round of decimals
```

```
[1] 313 104 104   35
```

7.
```
> obs = c(57,330,279,42)
> expP = c(.1,.4,.4,.1)
> chisq.test(obs, p = expP)
```

```
        Chi-squared test for given probabilities
```

```
data:  obs
X-squared = 22.2013, df = 3, p-value = 5.923e-05
```

No, these frequencies for individuals in the BMI categories are not consistent with the expected percentages ($\chi^2 = 22.2$, df $= 3$, p < 0.001).

9.
```
> a = chisq.test(obs,p= expP) # store return values
> round(a$exp,0) # round of decimals
```

```
[1]   71 283 283   71
```

CHAPTER 11

1.
```
> my.stats = function(x) {
+   my.mean = mean(x)
```

```
+   my.sd = sd(x)
+   my.SEM = my.sd/sqrt(length(x))
+   return(data.frame(my.mean,my.sd,my.SEM))
+ }
```

3.
```
> library(plotrix)
> A = c(8.8, 8.4, 7.9, 8.7, 9.1, 9.6)
> B = c(10.6, 9.8, 10.1, 8.4, 9.6, 10.2)
> C = c(11.6, 11.4, 9.1, 10.7, 14.9, 12.9)
> my.dat = data.frame(A,B,C)
> my.dat = stack(my.dat)
> names(my.dat) = c("Clotting","Drug")
> M = tapply(my.dat$Clotting,my.dat$Drug,mean)
> S = tapply(my.dat$Clotting,my.dat$Drug,sd)
> SEM = S/sqrt(6) # 6 counts per farm
> CI95 = qt(0.975,df=5)*SEM
> clot.aov = aov(my.dat$Clotting ~ my.dat$Drug)

> summary(clot.aov)

              Df Sum Sq Mean Sq F value  Pr(>F)
my.dat$Drug    2   28.2  14.102   8.778 0.00299 **
Residuals     15   24.1   1.606
---
Signif. codes: 0'***' 0.001'**' 0.01'*' 0.05'.'

> TukeyHSD(clot.aov)

  Tukey multiple comparisons of means
    95% family-wise confidence level

Fit: aov(formula = my.dat$Clotting ~ my.dat$Drug)

$`my.dat$Drug`
        diff        lwr        upr      p adj
B-A 1.033333 -0.86740707 2.934074 0.3598692
C-A 3.016667  1.11592626 4.917407 0.0024482
C-B 1.983333  0.08259293 3.884074 0.0403376

> a = barplot(M, ylim = c(0,15), cex.lab = 1.5,
+           xlab = "Drug",ylab = "Clotting Time (min)")
```

```
> abline(h=0)
> my.letters = c("a","a","b")
> plotCI(a,M,CI95,add = T, pch = NA)
> text(a,M+CI95,my.letters,pos = 3)
```

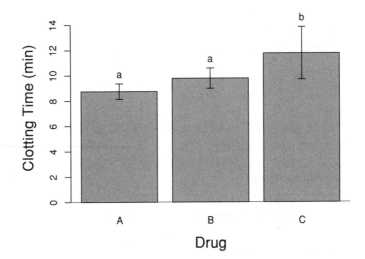

```
5. > speed = c(20,30,40,50,60)
   > mpg = c(24.1,28.1,33.1,33,31)
   > fit = nls(mpg ~ a*speed^2 + b*speed + c,
   +           start = list(a = -0.5, b = 1, c = 0))
   > a = coef(fit)[1]
   > b = coef(fit)[2]
   > c = coef(fit)[3]
   > # Set derivative equal to zero: 2*a*speed + b = 0
   > # Find speed, given a and b: speed = -b/(2*a)
   > cat("Maximum mpg at", round(-b/(2*a),1), "mph.\n")

   Maximum mpg at 47.7 mph.

7. > N0= 50
   > r = 0.82
   > K = 1000
   > curve((K * N0 * exp(r*x))/(K + N0 * (exp(r*x)-1)),0,10,
   +       ylab = "N", xlab = "Time", cex.lab = 1.5)
```

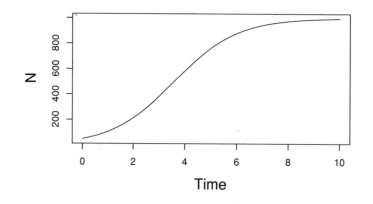

9. There are at least two ways to subset the data. Below I show these for the untreated sample (`untr`).

```
> attach(Puromycin)
> untr = Puromycin[state == "untreated",] # or use
> # dat = subset(Puromycin,state == "untreated")
> tr = Puromycin[state == "treated",]
> par(mfrow = c(1,2))
> plot(tr$conc,tr$rate, ylim = c(0,220), pch = 16,
+       xlab = "Concentration", ylab = "Rate",
+       main = "Treated")
> plot(untr$conc,untr$rate, ylim = c(0,220), pch = 16,
+       xlab = "Concentration", ylab = "Rate",
+       main = "Untreated")
> par(mfrow = c(1,1))
```

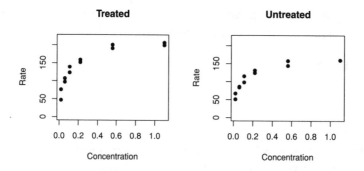

CHAPTER 12

1. This can be done easily without a `for` loop:

   ```
   > sum(1:100)
   ```

   ```
   [1] 5050
   ```

   but this, I guess, is cheating. To accomplish this the *hard* way, using a
   `for` loop, we can do the following:

   ```
   > the.sum = 0 # recall "sum" is a function, so avoid
   > for (i in 1:100) {
   +    the.sum = the.sum + i
   + }
   > cat("The sum is ",the.sum,"\n")
   ```

   ```
   The sum is   5050
   ```

3.
   ```
   > fib = c(0,1) # start the sequence
   > for (i in 3:10) { # do 8 more times
   +    fib[i] = sum(fib[i-1],fib[i-2])
   + }
   > par(mfrow = c(1,2))
   > plot(fib)
   > plot(log(fib))
   > par(mfrow = c(1,1))
   ```

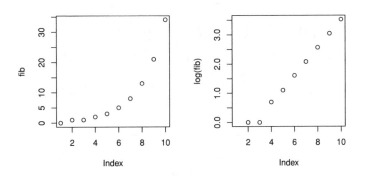

5. See the figure on the next page.

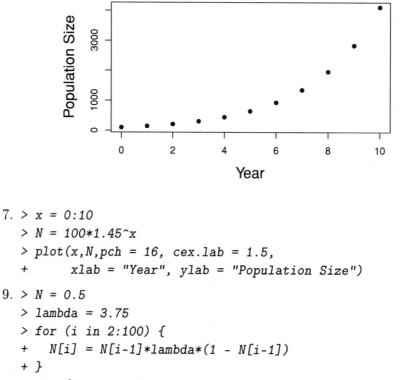

7. ```
> x = 0:10
> N = 100*1.45^x
> plot(x,N,pch = 16, cex.lab = 1.5,
+   xlab = "Year", ylab = "Population Size")
```

9. ```
> N = 0.5
> lambda = 3.75
> for (i in 2:100) {
+   N[i] = N[i-1]*lambda*(1 - N[i-1])
+ }
> plot(N,type = "l", xlab = "Time Step", ylim = c(0,1),
+       cex.lab = 1.5)
```

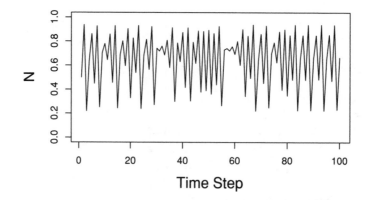

BIBLIOGRAPHY

Adler, J., 2012. R in a nutshell: A desktop quick reference. O'Reilly.

Chang, W., 2013. R graphics cookbook. O'Reilly.

Crawley, M., 2012. The R book. Wiley.

Focht, D., C. Spicer, and M. Fairchok, 2002. The efficacy of duct tape vs. cryotherapy in the treatment of verruca vulgaris (the common wart). *Archives of Pediatrics & Adolescent Medicine* **156**:971–974.

Gotelli, N. and A. Ellison, 2012. A primer of ecological statistics, 2nd ed. Sinauer Associates, Inc.

Hartl, D. L. and D. J. Fairbanks, 2007. Mud sticks: On the alleged falsification of Mendel's data. *Genetics* **175**:975–979.

Hartvigsen, G., 2011. Using R to build and assess network models in biology. *Mathematical Modeling of Natural Phenomena* **6**:61–75.

Marshall, W., H. Qin, M. Brenni, and J. Rosenbaum, 2005. Flagellar length control system: Testing a simple model based on intraflagellar transport and turnover. *Molecular Biology of the Cell* **16**:270–278.

Matloff, R., 2011. The art of R programming. No Starch Press.

Mendel, G., 1866. Versuche über pflanzen-hybriden. Verhandlungen des naturforschenden vereines. *Abh. Brünn* **4**:3–47.

Meys, J. and A. de Vries, 2012. R for dummies. For Dummies.

Nelson, W., O. Bjornstad, and T. Yamanaka, 2013. Data from: Recurrent insect outbreaks caused by temperature-driven changes in system stability. *Dryad Digital Repository* **341**:796–799.

Pruim, R., 2011. Foundations and applications of statistics: An introduction using R. American Mathematical Society.

Silver, N., 2012. The signal and the noise: Why so many predictions fail – but some don't. Penguin Press HC.

Snedecor, G. and W. Cochran, 1989. Statistical methods, 8th ed. Blackwell Publishing Professional.

Templeton, C., E. Greene, and K. Davis, 2005. Allometry of alarm calls: Black-capped chickadees encode information about predator size. *Science* **308**:1934–1937.

Thomas, R., M. Fellowes, and P. Baker, 2012. Spatio-temporal variation in predation by urban domestic cats (*Felis catus*) and the acceptability of possible management actions in the UK. *PLoS ONE* **7**:e49369.

Ugarte, M., A. Militino, and A. Arnholt, 2008. Probability and statistics with R. Chapman and Hall/CRC.

Venables, W. and D. Smith, 2009. An introduction to R. Network Theory Ltd.

Ware, W., J. Ferron, and B. M. Miller, 2012. Introductory statistics: A conceptual approach using R. Routledge.

Wickham, H., 2009. ggplot2: Elegant graphics for data analysis. Springer.

Zar, J., 2009. Biostatistical analysis. Prentice Hall.

Zuur, A., E. N. Ieno, and E. Meesters, 2009. A beginner's guide to R. Springer.

INDEX